POLLUTION SCIENCE, TECHNOLOGY AND ABATEMENT

CARBON DIOXIDE EMISSIONS

PAST, PRESENT AND FUTURE PERSPECTIVES

POLLUTION SCIENCE, TECHNOLOGY AND ABATEMENT

Additional books and e-books in this series can be found on Nova's website under the Series tab.

POLLUTION SCIENCE, TECHNOLOGY AND ABATEMENT

CARBON DIOXIDE EMISSIONS

PAST, PRESENT AND FUTURE PERSPECTIVES

ASIA SANTANA
EDITOR

Copyright © 2020 by Nova Science Publishers, Inc.

All rights reserved. No part of this book may be reproduced, stored in a retrieval system or transmitted in any form or by any means: electronic, electrostatic, magnetic, tape, mechanical photocopying, recording or otherwise without the written permission of the Publisher.

We have partnered with Copyright Clearance Center to make it easy for you to obtain permissions to reuse content from this publication. Simply navigate to this publication's page on Nova's website and locate the "Get Permission" button below the title description. This button is linked directly to the title's permission page on copyright.com. Alternatively, you can visit copyright.com and search by title, ISBN, or ISSN.

For further questions about using the service on copyright.com, please contact:
Copyright Clearance Center
Phone: +1-(978) 750-8400   Fax: +1-(978) 750-4470   E-mail: info@copyright.com

NOTICE TO THE READER

The Publisher has taken reasonable care in the preparation of this book, but makes no expressed or implied warranty of any kind and assumes no responsibility for any errors or omissions. No liability is assumed for incidental or consequential damages in connection with or arising out of information contained in this book. The Publisher shall not be liable for any special, consequential, or exemplary damages resulting, in whole or in part, from the readers' use of, or reliance upon, this material. Any parts of this book based on government reports are so indicated and copyright is claimed for those parts to the extent applicable to compilations of such works.

Independent verification should be sought for any data, advice or recommendations contained in this book. In addition, no responsibility is assumed by the Publisher for any injury and/or damage to persons or property arising from any methods, products, instructions, ideas or otherwise contained in this publication.

This publication is designed to provide accurate and authoritative information with regard to the subject matter covered herein. It is sold with the clear understanding that the Publisher is not engaged in rendering legal or any other professional services. If legal or any other expert assistance is required, the services of a competent person should be sought. FROM A DECLARATION OF PARTICIPANTS JOINTLY ADOPTED BY A COMMITTEE OF THE AMERICAN BAR ASSOCIATION AND A COMMITTEE OF PUBLISHERS.

Additional color graphics may be available in the e-book version of this book.

Library of Congress Cataloging-in-Publication Data

ISBN: 978-1-53617-763-3

Published by Nova Science Publishers, Inc. † New York

Contents

Preface		**vii**
Chapter 1	Recent Process Technologies for CO_2 Capture and Removal from Flue Gases *João F. Gomes, Samuel P. Santos, Ana P. Duarte and João C. Bordado*	**1**
Chapter 2	Soil CO_2 Emission in Brazil *Gabriel Ribeiro Castellano*	**41**
Chapter 3	Brazil at COP21: Challenges to Achieve Carbon Emission Reduction Targets *Marcelo Silva Sthel, Marcenilda Amorim Lima and Fernanda Gomes Linhares*	**67**
Chapter 4	Estimates of the Inflation Effect of a Global Carbon Price on Consumer, Investment, Export and Import Prices *Fredrik N. G. Andersson*	**87**
Index		**113**

PREFACE

CO2 capture from gaseous effluents is one of the great challenges faced by chemical and environmental engineering, as the increase of CO2 levels in the Earth's atmosphere might be responsible for dramatic climate changes. This compilation begins by presenting the recent developments in studies focusing on the optimization of CO2 capture using amine solutions.

The authors assess the effects of land use change on soil carbon flux in Brazil, in addition to contributing to the body of knowledge about carbon stock balance in tropical and subtropical domains.

The authors also assess whether it will it be possible to fulfill the Brazilian Paris agreement goals if the Amazon deforestation increase continues.

The potential inflation effects of a global carbon price on consumer prices, investment prices, export prices, and import prices are explored, estimating the effects under various scenarios.

Chapter 1 - CO_2 capture from gaseous effluents is one of the great challenges faced by chemical and environmental engineering, as the increase of CO_2 levels in the Earth atmosphere might be responsible for dramatic climate changes. From the existing capture technologies, the only proven and mature technology is chemical absorption using aqueous amine solutions. However, bearing in mind that this process is somewhat expensive, it is important to choose the most efficient and, at the same time, the least expensive solvents. This chapter presents recent development studies towards the optimization of CO_2 capture using amine solutions. For this purpose, a pilot test facility was assembled to study the absorption behaviour of aqueous amine solutions and perform bench-marking studies. Using othe similar, but large equipment it was also possible to perform scale-up studies. Finally, the description of a method for simultaneous removal of CO_2, SOx and NOx from industrial flue gases is presented.

Chapter 2 - The signficant increase in atmospheric CO_2 in the last century is primarily due to fossil fuel combustion (36.6 Gt of CO_2 in 2018) and land-use change/deforestation (5.5 Gt of CO_2 per year from 2009 to 2018). In Brazil, agricultural activities account for 22% of total CO_2 emission. Land use change, the main cause of CO_2 emission in the country, accounts for 51%. These changes occur mainly in forests and savannas, because their soil and climate conditions are ideal for high-yield agricultural production. Changes in land cover significantly alter physical, biological, and chemical characteristics of soils. Soil CO_2 emissions (FCO_2) is a result of physical and biochemical processes that determine CO_2 production and transport from soil to atmosphere. CO_2 production is related to microorganism activity and plant root respiration, whereas CO_2 transport is associated to the physical structure of the soil, especially its porosity, which affects soil gas flux. Based on pooled data from FCO_2 research carried out in Brazil from 1990 to 2019 with IRGA (infra-red gas analyzer), this study aims to assess the effects of land use change on soil carbon flux in Brazil, in addition to contributing to the body of knowledge about carbon stock balance in tropical and subtropical domains. A bibliographical review was conducted and data from research done in the Amazon Forest, Atlantic Forest, Cerrado (South America savanna), and agricultural crops were pooled. FCO_2 in the Amazon Forest ranged from 3.2 to 6.4 $\mu mol\ CO_2\ m^{-2} s^{-1}$; several studies reported a significant linear correlation ($p < 0.05$) between FCO_2 and soil moisture. FCO_2 in the Atlantic Forest ranged from 0.51 to 3.86 $\mu mol\ CO_2\ m^{-2}\ s^{-1}$, indicating a significant linear correlation with soil moisture ($r = 0.55$, $p < 0.0001$). FCO_2 in the Cerrado was 2.55 μmol and 0.86 $\mu mol\ CO_2\ m^{-2}\ s^{-1}\ CO_2\ m^{-2}\ s^{-1}$ in the rainy and dry seasons, respectively. In agricultural crops in Brazil's southeast (São Paulo State), FCO_2 ranged from 1.19 up 5.3 $mol\ CO_2\ m^{-2}\ s^{-1}$. Most of these studies were conducted in sugarcane plantations using spatial variability analysis, while those carried out in natural areas focused on temporal variability. Studies in Brazil indicate that soil moisture has the most important temporal influence on FCO_2 in forests and savannas. In agricultural crops, FCO_2 values are affected by soil and agricultural management practices. The implications of the land cover change in Brazil need to be discussed.

Chapter 3 - In December 2015, the United Nations Framework Convention on Climate Change (UNFCCC) held the Conference of the Parties (COP21). Brazil has committed to reduce Greenhouse Gasses (GHGs) emissions in 37% by 2025, and 43% by 2030, based on 2005 levels. The current Brazilian policy for deforestation control in the Amazon region is in line with the emission-reducing promises made at COP21. Brazil has programs for reducing GHGs via Intended Nationally Determined Contributions (INDC), by transforming the land use and the forestry sector. The country has made a pledge to adopt additional measures in this sector so as to achieve the goals agreed in Paris, such as: strengthening compliance with the Brazilian Forestry Code at national, state and city levels; supporting policies to achieve zero illegal deforestation in the Brazilian Amazon by 2030; restoring and reforesting 12 million hectares of forests by 2030, for multiple uses; and scaling up sustainable native forest management systems through georeferencing and traceability applicable to native forest management, in order to discourage illegal and unsustainable practices. In January 2019, a new government took position in Brazil by promising to change the Brazilian environmental policy. In the following August, the National Space Research Institute (INPE) released data on Amazon deforestation, which indicated a considerable growth when compared with 2018, thus contradicting the additional measures proposed by the Brazilian government. This fact created an international crisis between Brazil and the G7 summit European countries. In this way, in case the Amazon deforestation increase continues, will it be possible to fulfill the Brazilian Paris agreement goals?

Chapter 4 - This chapter considers the potential inflation effects of a global carbon price on consumer prices, investment prices, export prices, and import prices. The authors estimate the effects under three different scenarios. The results clearly indicate that the inflation effects in developed countries of a 100 USD/ton carbon price are small. For developing countries, the inflation effect is larger—potentially too large for it to be politically feasible to introduce a global carbon price. However, a simple adjustment of the price based on the price level in each country equalizes the inflation effects across all countries. In light of this kind of adjustment, a global carbon price is more likely to be implemented.

In: Carbon Dioxide Emissions
Editor: Asia Santana
ISBN: 978-1-53617-763-3
© 2020 Nova Science Publishers, Inc.

Chapter 1

RECENT PROCESS TECHNOLOGIES FOR CO_2 CAPTURE AND REMOVAL FROM FLUE GASES

João F. Gomes[1,2,*]*, PhD, Samuel P. Santos*[2]*,
Ana P. Duarte*[3] *PhD and João C. Bordado*[2] *PhD*

[1]Área Departamental de Engenharia Química,
ISEL – Instituto Superior de Engenharia de Lisboa, Lisboa, Portugal
[2]CERENA, IST – Instituto Superior Técnico, Lisboa, Portugal
[3]Universidade Atlântica, Barcarena, Portugal

ABSTRACT

CO_2 capture from gaseous effluents is one of the great challenges faced by chemical and environmental engineering, as the increase of CO_2 levels in the Earth atmosphere might be responsible for dramatic climate changes. From the existing capture technologies, the only proven and mature technology is chemical absorption using aqueous amine solutions. However, bearing in mind that this process is somewhat expensive, it is important to choose the most efficient and, at the same time, the least expensive solvents. This chapter presents recent development studies towards the optimization of CO_2 capture using amine solutions.

---

[*] Corresponding Author's E-mail: jgomes@deq.isel.ipl.pt.

For this purpose, a pilot test facility was assembled to study the absorption behaviour of aqueous amine solutions and perform benchmarking studies. Using othe similar, but large equipment it was also possible to perform scale-up studies. Finally, the description of a method for simultaneous removal of CO_2, SOx and NOx from industrial flue gases is presented.

Keywords: carbon dioxide, chemical absorption, aqueous amine solution, scale-up, simultaneous removal

INTRODUCTION

CO_2 capture from gaseous effluents is, nowadays, one of the great challenges faced by chemical and environmental engineering, as the increase of CO_2 levels in the Earth atmosphere is endangering the support of living species in this planet and might also be responsible for dramatic climate changes [1]. From the existing capture technologies, the only proven and mature technology is, currently, chemical absorption using aqueous amine solutions [2-4]. This is due to the fact that gas absorption has been used extensively, since the 50s, for treatment of natural gas thus removing sour gases such as CO_2 and H_2S. By that time, the main reason for treating those gases was related with obtaining more pure gaseous streams stripped from these acidified species that are bound to create corrosion problems and, also, decrease the gas heating value [5]. Also, further economic benefits could be obtained by obtention of purified CO_2 [6]. Although this is a proven process, within the natural gas industry, other problems take place when this technology is to be applied to the treatment of gaseous effluents from power plants. In these cases: i) the gas temperature is usually high, around 150 °C; ii) the pressure is low, usually slightly higher than atmospheric pressure; iii) apart from CO_2 the gaseous streams also contains other acid contaminants such as SO_2, NO_x and fine particulate; iv) has a low CO_2 content. The CO_2 content depends mainly on the nature of the fuel burned in the power plant ranging from 3% in a natural gas fired power plant to 15% in a coal fired power plant [7].

All these characteristics do not favour chemical absorption, particularly bearing in mind the high flow of gaseous effluent to be treated, which usually ranges hundreds of millions of kg/h. The low CO_2 content does not originate a high CO_2 partial pressure needed to create a high concentration gradient which is a prime condition to increase mass transfer [8].

Apart from this, there are still other operational problems to overcome, namely those resulting from solvent degradation, precipitation, corrosion and foaming. Another problem is related with the high energy consumption of the whole process: the capture process includes a first column where absorption takes place by contacting, in countercurrent, with the amine solution and, in a second column, the process is reversed releasing the previously absorbed CO_2, and regenerating the amine solution so that it can be used again to promote absorption in the first column [9]. It should be noted that the energy consumed in the second column is inversely proportional to the reactivity of the amine in terms of CO_2 absorption. In fact, for better CO_2 capture it would be desired to use a very reactive amine, such as a primary amine, whereas for amine regeneration, thus releasing CO_2, it would be more convenient to use less reactive amines, such as secondary or tertiary amines, resulting in lower energy consumption in the second column [10]. As shown in Figure 1, the chemical absorption mechanism involves the reaction of CO_2 with the amine, originating ammonium carbamate, which, in aqueous phase is converted to bicarbonate, thus fixing CO_2 [11].

It has been shown previously [12-14] that relationships exist between the amine structure and the activity and capacity for CO_2 absorption. Apparently, the introduction of amine substituents at the α-carbon creates a carbamate instability, which causes the hydrolysis to go faster, thus increasing the amount of bicarbonate, allowing for higher CO_2 loadings [12]. To obtain a better understanding of the structure-activity relationship it is necessary to perform solvent screening experiments, in order to investigate the effect of variables such as chain length, increase in number of functional groups, side chain at α-carbon position, alkyl group position in cyclic amine and side effect of cyclic amine with different functional groups. The description of these effects, in a quantitative way, on the initial rate of absorption for CO_2, as well as the capacity of various solvents for CO_2 absorption will greatly benefit on designing more efficient absorption systems for CO_2 capture from flue gases [15].

STUDYING THE ABSORBING BEHAVIOUR OF AMINES

To study the absorbing behaviour of amine solutions, a pilot unit was used, and its flowsheet is shown in Figure 2. This includes an absorption column, as well as a stripping column, a heat exchanger between the two columns, a reboiler for the stripping column, pumping systems, surge tanks and all necessary instrumentation and control systems. The design features of the absorption column are shown in Table 1.

In order to simulate the expected CO_2 concentration in the gaseous streams, CO_2 from a cylinder (99.99% Air Liquide) (3) could be mixed with compressed air from a compressor (4). (in this particular study, tests were made using pure CO_2) and the gaseous stream leaving the top of the absorption column is connected to a CO_2 analyser (WITT GasTechnik) (8), so that the amount of CO_2 existing in that stream could be measured. An aspect of the pilot unit is presented in Figure 3.

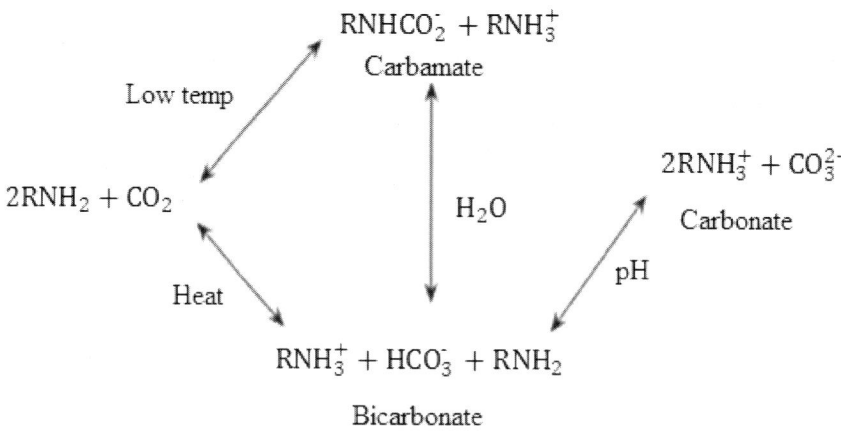

Figure 1. Proposed reaction mechanism for CO_2 absorption [11].

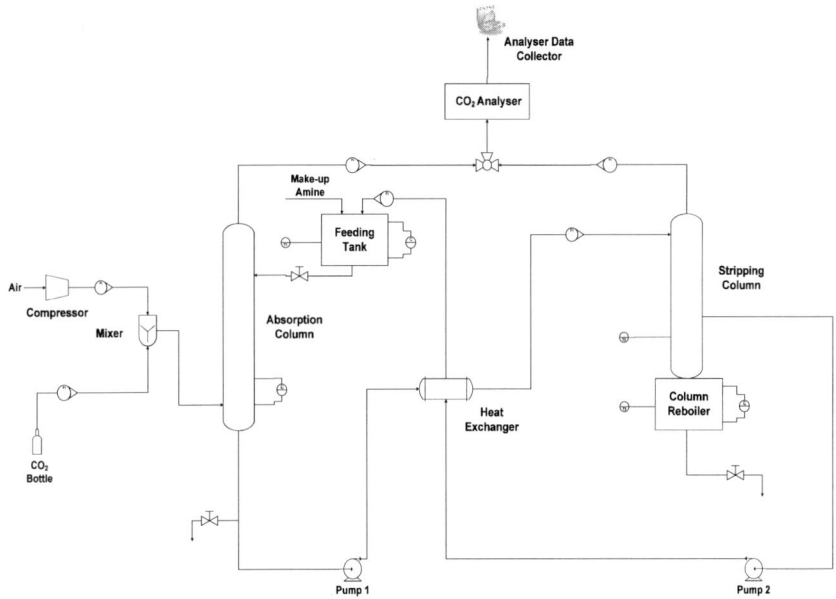

Figure 2. Flowsheet of pilot installation.

Table 1. Design features of the absorption column

Height (cm)	150
Outside diameter (cm)	5.05
Thickness (cm)	0.2
Operating pressure range (atm)	1-3
Construction material	Stainless steel AISI 316
Packing material	Mild steel wool ASSOLAN

In order to study the loading capacity of the studied amines in the pilot plant unit, aqueous amine solutions, referenced hereafter, were prepared from p.a. reagents (VWR) having a concentration of 10% (w/w). The amount of amine needed for the solution, was calculated considering that the feeding tank in the pilot unit had a volume of 7 L. Then, for each test, each solution was placed inside the feeding tank, after which. the respective control valve was opened in order to enable the solution to flow into the absorption column. The flow rate of the aqueous amine solutions inside the absorption column was 24 L/h and the CO_2 stream entered the absorption column from the bottom with a flow rate of 20 mL/min. Then, the first dosing pump was turned on, in order to allow the CO_2 "rich" aqueous amine solution, which left the absorption column, to be directed into the stripping column. Afterwards, the second dosing pump was turned on, in order to allow the aqueous amine solution leaving the stripping column to be directed to the feeding tank, thus closing the absorption cycle. Temperatures inside the stripping column are shown in Table 2, for each amine solution tested.

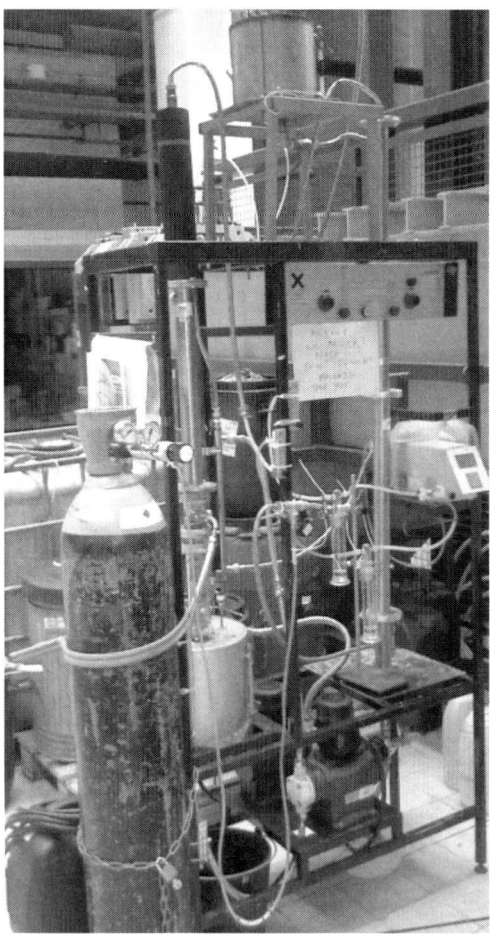

Figure 3. Pilot unit.

Table 2. Temperatures inside the stripping column for each amine solution tested

Aqueous Amine Solution	Temperature inside stripping column (°C)
MEA	69
EDA	89
Piperazine	88
Bis(2-ethylhexyl)amine	30
Triethylamine	36
MDEA	50
DEA	60

Every 30 minutes, a sample of 20 mL was collected from the absorption column, in order to ascertain whether the solution was or was not saturated, through a sampling valve. The sample was then analyzed by the $BaCO_3$ precipitation method. To use this method, the procedures referred by Santos [16] and Li et al. [17] were followed: after the amine solution had been saturated with CO_2, a sample of 20 mL was collected from the pilot unit. Then, a solution of sodium hydroxide (NaOH) 1.0 M, previously prepared from commercial sodium hydroxide (Solvay), was added in excess to the sample so that the dissolved CO_2 was converted to non-volatile ionic species. Then, a solution of barium chloride dihydrate ($BaCl_2\ 2H_2O$) 1.0 M, prepared previously from barium chloride dehydrate (Panreac) was added in excess. The solution was well stirred to ensure that all the CO_2 was absorbed, and precipitated as barium carbonate ($BaCO_3$). Afterwards, the solution containing the precipitate was filtrated, dried and weighted. The amount of precipitate was used to calculate the CO_2 loading capacity, as CO_2 moles per mol of amine. The equations used in the calculations are as follows:

$$w_{amine} = w_{sample} \times \%_w \qquad (1)$$

where:

w_{amine} – corresponds to the amine weight (g);

w_{sample} – corresponds to the weight of the sample collected from the pilot unit (g);

%$_w$ – corresponds to the concentration of aqueous amine solution.

$$n_{amine} = \frac{w_{amine}}{MW_{amine}} \quad (2)$$

where:

n_{amine} – corresponds to the number of moles of amine (mol amine);
w_{amine} – corresponds to the amine weight calculated before in Eq. 1;
MW_{amine} – corresponds to the molecular weight of the amine (g/mol).

$$n_{CO_2} = \frac{m_{precipitate}}{MW_{BaCO_3}} \quad (3)$$

where:

n_{CO2} – corresponds to the number of moles of the obtained CO_2 (mol CO_2);
$m_{precipitate}$ – corresponds to the weight of the obtained precipitate (g);
MW_{BaCO3} – corresponds to the molecular weight of $BaCO_3$ (g/mol).

$$\alpha = \frac{n_{CO_2}}{n_{amine}} \quad (4)$$

where:

α – corresponds to the CO_2 loading capacity of the aqueous amine solution (mol CO_2/mol amine);
n_{CO2} – corresponds to the number of moles of the obtained CO_2 calculated before as in Eq. 3 (mol CO_2);
n_{amine} – corresponds to the number of moles of the amine calculated before as in Eq. 2 (mol amine)

Tests performed in the pilot unit comprised amine solutions already indicated as good absorbents during preliminary laboratory tests by Santos [16], as well as other promising amine absorbing solutions, that were prepared from pro-analysis Merck reagents in 10% concentration by weight in water: monoethanolamine (MEA), diethanolamine (DEA), diethylamine, ethylenediamine (EDA), N-methyldiethanolamine (MDEA) and piperazine

(PZ). In order to determine the CO_2 loading capacity by the aqueous amine solutions, the precipitation method was used. The amount of formed precipitate from the addition of $BaCl_2\ 2H_2O$ was used to calculate the CO_2 loading capacity, in terms of moles of CO_2 per moles of amine [17].

When analyzing these results it can be noted that the CO_2 loading capacity for each aqueous amine solution increases, with some fluctuations, over time. In other words, the aqueous amine solutions initially act like a "lean" solvent because they had not absorbed any CO_2 yet. As the contact time (in the pilot unit) between the solutions and the CO_2 increases the amount of absorbed CO_2 also increases. It can also be noted that, using the precipitation method, the aqueous amine solution that showed a higher CO_2 loading capacity was diethylamine. This amine can absorb 0.492 mol of CO_2 per mol of amine, against the 0.409 mol of CO_2 per mol of amine obtained by MEA, typically considered as the benchmark solvent to which alternative solvents are to be compared [18]. The loading capacity achieved by PZ at the end of the test was very close to what MEA presented. This means that PZ could be a good alternative solvent for CO_2 absorption. EDA was the amine that showed the worst loading capacity (0.321 mol of CO_2 per mol of amine) [19]. It can also be noted that the obtained results for MDEA and DEA show that these amines have a higher CO_2 loading capacity than the other amine solutions used in this work. The obtained results are also shown in Figure 4.

By analyzing Figure 4 it can be observed that diethylamine, a secondary alkylamine, showed the highest CO_2 loading capacity of the four amine solutions that were tested, although in the first sample taken, at the moment of 30 minutes, it was the one that had absorbed the lowest amount of CO_2. The curve that shows the loading capacity of PZ over absorption time has, at the moment of 120 minutes, a strange fluctuation. The reason why, at this particular instant the loading capacity of PZ decreased was due to the fact that the CO_2 flow inside the absorption column was higher than the flow of the PZ solution. This created a problem, as the CO_2 flow inside the column was such that it did not allow the PZ solution to flow downwards, thus not allowing the reaction to take place normally. Besides that, when comparing with MEA, which is the benchmark solvent, PZ showed, in some moments, a higher loading capacity. In the case of a sterically hindered compound such as PZ, the chemical reaction is particularly important, as the presence of the methyl group significantly reduces the stability of the carbamate bond, resulting in the preferred formation of the bicarbonate, and thus leading to the particularly high loading capacity of this solvent. For EDA, its loading capacity over absorption time curve indicates that, despite the result at 30 minutes, this was the amine solution that showed the worst results through the whole duration of the absorption test [19]. The fluctuation that is seen in the absorption curves of the four amine solutions is probably due to experimental errors associated with the precipitation method, more specifically the precipitate filtration and drying processes. As can be also seen in Figure 4, MDEA and DEA showed a higher loading capacity throughout the absorption tests and needed a shorter absorption time (180 minutes) to be saturated. In the case of MDEA, as it is a tertiary amine and it does not react directly with CO_2, but acts as a base, catalyzing the hydration of CO_2. This means that it has an equilibrium CO_2 loading capacity nearly of 1.0 mol CO_2/mol amine and, thus, the obtained results are consistent with what was expected [20]. For DEA, as it is a secondary amine, based on stoichiometry, the expected CO_2 loading capacity was 0.5 mol CO_2/mol amine. However, the obtained results showed that the loading capacity of DEA was close to 1.0 mol CO_2/mol amine [20].

In order to make a cost analysis for the studies performed in the pilot unit the market value of each amine was obtained and related to the CO_2 loading capacity, as presented in Figure 5.

The amine prices shown in Figure 5 correspond to the lowest market value that was found in Portugal. The prices are for the pure amines. Due to the fact that, in the performed tests, the amine solutions had the same concentration (10% (w/w)) these prices do not need to take into account the dilution factors. MEA shows the second highest loading capacity (0.409 mol of CO_2 per mol of amine) and the lowest price per liter (30.50 €/L). As it is the benchmark solvent for the chemical absorption of CO_2, the other amine solutions were compared to it. PZ is the one that showed a loading capacity close to MEA (0.395 mol of CO_2 per mol of amine) but it has the highest price per liter of all the four amines tested in this study (68.70 €/L).

Diethylamine was the one that showed the highest loading capacity (0.492 mol of CO_2 per mol of amine), but its price per liter was very close to the price for PZ (66.40 €/L). EDA had the lowest loading capacity (0.321 mol of CO_2 per mol of amine) and its price per liter is close to the price for MEA (31.70 €/L).

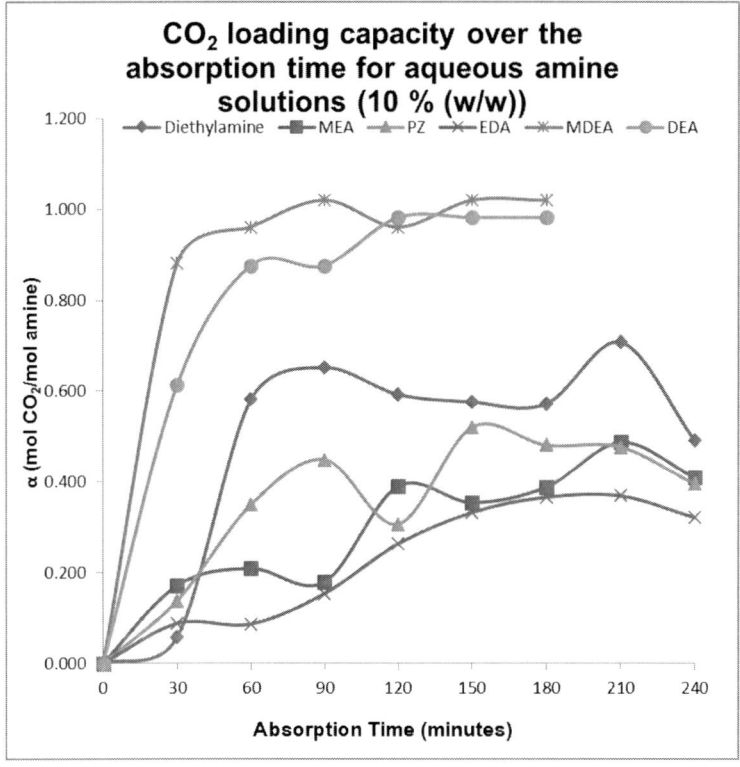

Figure 4. Comparison of the obtained results for CO_2 loading capacity over the absorption time for all the tested aqueous amine solutions (10% (w/w)).

However, it can be also concluded that the amine with the best cost/loading capacity and, therefore, the best choice for the chemical absorption of CO_2 would be DEA, because it shows a high loading capacity (0.982 mol of CO_2 per mol of amine) and its price it is even lower than MEA (25.70 €/L).

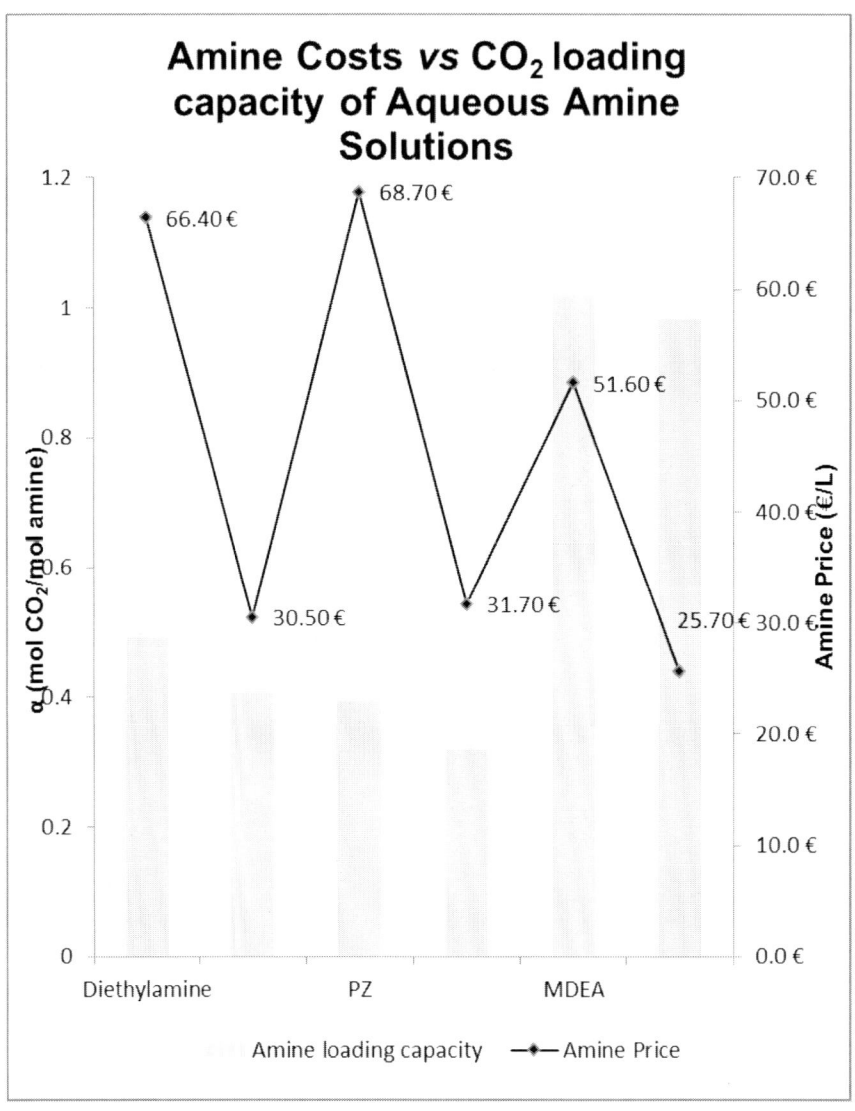

Figure 5. Amine Price versus CO_2 loading capacity of Aqueous Amine Solutions.

In the studies performed in the pilot unit, six amine solutions were tested: diethylamine, MEA, PZ, ED, MDEA and DEA. These amines were used to prepare aqueous amine solutions with a concentration of 10% (w/w). These tests were performed with an absorption time ranging from 180 to 240 minutes for all the amine solutions.

When analyzing the obtained results, it can be concluded that the amine solution that showed the best results, with the highest CO_2 loading capacity (0.492 mol of CO_2 per mol of amine) was the diethylamine aqueous solutions. The MEA aqueous solution that in this study was considered as the benchmark amine aqueous solution for the chemical absorption of CO_2, showed the second highest loading capacity, followed very closely by the PZ aqueous solution. The EDA aqueous solution showed the lowest loading capacity, which means that this amine solution could not be the best alternative solvent to replace MEA as a solvent of choice in CO_2 chemical absorption.

A cost analysis was also made in order to see which one of the amines was the most economical choice of solvent for the chemical absorption of CO_2. Considering the amines tested in this work, DEA is the one that turned out to be the most economical choice, as it showed a higher CO_2 loading capacity (0.982 mol of CO_2 per mol of amine), although it was not the highest (MDEA had a loading capacity of 1.020 mol of CO_2 per mol of amine), and the lowest price per liter (25.70 €/L), even when compared with MEA, the benchmark solvent, exhibiting a price per liter of 30.50 €/L.

SCALE-UP EFFECTS OF CO_2 CAPTURE BY METHYLDIETHANOLAMINE (MDEA) SOLUTIONS IN TERMS OF LOADING CAPACITY

For this study, on the behavior of MDEA solutions regarding CO_2 absorption, three different experimental scales were used in order to collect consistent data, a laboratory scale, where a 500-mL three-necked flask was used, and two pilot-units, whose dimensions are shown in Table 3. All tests were performed at 1 atm and 20°C.

Table 3. Dimensions of the absorption column used for the different scales

Scale	Column Height (dm)	Column Inner Diameter (dm)	Column Volume (L)	CO_2 Flowrate (L/h)	Aqueous Amine Solution Flow (L/h)
Pilot-unit	15	0.505	3	1.2	24
Large pilot-unit	110	0.750	62	1200	40

Technique for CO_2 Dosing

For CO_2 dosing, samples were collected from the absorption column through a sampling valve, in order to ascertain whether the solution was or was not saturated. The sample was then analyzed by the $BaCO_3$ precipitation method. To use this method, the procedures described by Li et al. [21] and Santos [22] were followed: After the amine solution had been saturated with CO_2, a sample of 20 mL was collected from the pilot unit. Then, a solution of sodium hydroxide (NaOH) 1.0 M, previously prepared from commercial sodium hydroxide (Solvay), was added in excess to the sample so that the dissolved CO_2 was converted to non-volatile ionic species. Then, a solution of barium chloride dihydrate ($BaCl_2 \cdot 2H_2O$) 1.0 M, prepared previously from barium chloride dehydrate (Panreac Quimica Slu, Barcelona, Spain), was added in excess. The solution was well stirred to ensure that all the CO_2 was absorbed, and was precipitated as barium carbonate ($BaCO_3$). Afterwards, the solution containing the precipitate was filtrated, dried, and weighted. The amount of precipitate was used to calculate the CO_2 loading capacity as CO_2 moles per mol of amine. The equations used in the calculations were described in the previous section as equations 1 to 4.

Laboratory Scale Trials

In a first stage, a laboratory-scale trial was performed to assess the maximum loading capacity of the prepared MDEA solution. A MDEA (Merck) aqueous solution (20 wt%, with distilled water), was introduced in a flask. Initially, the system was purged with N_2 for several minutes in order to remove any gaseous contaminants which could be possibly present. Then, CO_2 (99.99% Air Liquide) was carefully introduced into the flask to promote the saturation of the solution, as shown in Figure 6 [23]. Saturation was considered complete throughout the observation of the evolution of CO_2 bubbles into the aqueous amine solution.

In order to determine the CO_2 loading capacity of each aqueous amine solution under test until saturation, the method of barium chloride precipitation described in the previous section was used.

Pilot Unit Scale Trials

In the following stage, in order to study the absorbing behaviour of MDEA in a carbon capture and separation system (CCS), two pilot units were used. Both units include an absorption column, as well as a stripping column, a heat exchanger between the two columns, a reboiler for the stripping column, pumping systems, surge tanks, and all necessary instrumentation and control systems.

The second stage of this study was performed using a smaller pilot unit in order to assess whether the results obtained in the trials of the first stage were still valid at a larger scale. The flowsheet of this unit is presented in Figure 2, and an actual picture was already shown in Figure 3.

To simulate the expected CO_2 concentration in the gaseous streams, CO_2 from a cylinder (99.99% Air Liquide, Paris, France) was mixed with air from a compressor. Then, the gaseous solution entered the absorption column and contacted with the aqueous amine solution coming from the feed tank, and the chemical absorption of CO_2 was able to take place. The gaseous stream leaving the top of the absorption column was connected to a CO_2 analyzer

(Witt Gastechnik GmbH & Co KG, Witten, Germany) so that the amount of CO_2 existing in that stream could be measured. The cold liquid stream leaving the bottom of the absorption column flowed through a dosing pump and then went through a heat exchanger where it was heated. The liquid stream reached the top of the stripping column where the CO_2 release occurs. The gaseous stream, leaving the top of the stripping column, was connected to the stream leaving the top of the absorption column to the CO_2 meter so that the amount of CO_2 in that stream could be also measured. The liquid stream leaving the stripping column at high temperature flowed through a dosing pump and went through the heat exchanger, where it heated the stream leaving the bottom of the absorption column. This stream was cooled down in the heat exchanger and then went back into the feeding tank, where amine make-up was completed. Then, it entered the top of the absorption column, thus closing the cycle. The absorption column was filled with Berl saddles packing. Samples were collected just after the absorption column.

In order to study the loading capacity of the amine under test, aqueous solutions 20 wt% were prepared and placed in the feeding tank. CO_2 was then circulated jointly with the aqueous amine solution, and samples were collected every 30 min until the amine saturation was achieved. The amount of formed precipitate from the addition of $BaCl_2 \cdot 2H_2O$ was used to calculate the CO_2 loading capacity in terms of moles of CO_2 per moles of amine, following the procedure previously described.

In the third stage, tests were performed in a larger pilot unit constructed in accordance with the same flowsheet, but using larger columns and higher flowrates. Moreover, due to the size of pilot unit, a closed circuit television (CCTV) system was installed so that, with the help of the monitors placed throughtout the pilot unit, it was possible to observe what was happening during the tests. A Berl saddles packing was placed inside the absorption column in order to enable better contact between the aqueous MDEA solution and the CO_2 stream. CO_2 flows jointly with the aqueous amine solution, and samples were collected every 30 min until the amine saturation was achieved, again after the absorption column. In order to determine the CO_2 loading capacity by the aqueous MDEA solutions, the precipitation method was used again. A photo of this pilot unit was shown in Figure 7. Both units were operated so that the system achieves nearly a steady state.

Figure 6. Illustration of experimental tests carried out to study the absorption of CO_2 from aqueous amines solutions [23].

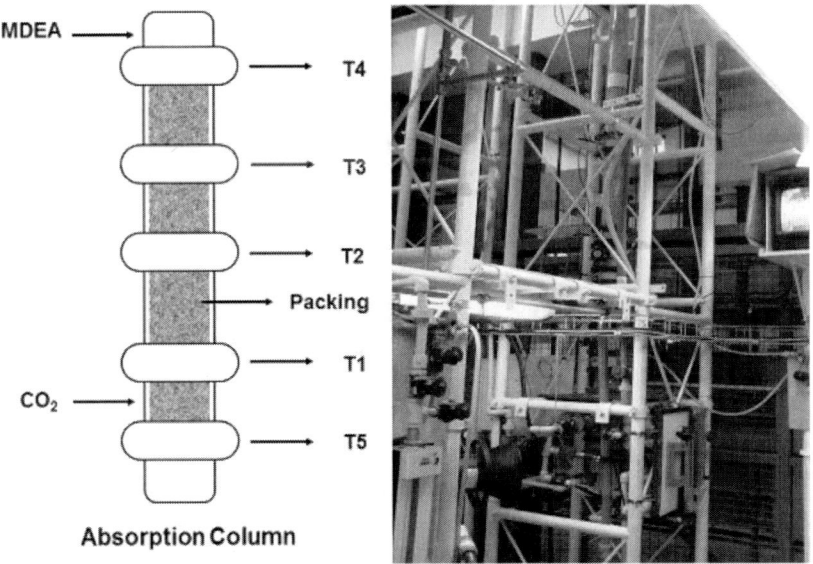

Figure 7. Large pilot unit and absorption column detail.

Results from Laboratory Scale Trials

In the laboratory scale trial performed using a 500 mL three-necked flask, the maximum loading capacity of the prepared MDEA solution was assessed. CO_2 was bubbled inside the aqueous MDEA solution until it was saturated. This process took around four hours to be achieved.

Using the precipitation method, as described before, the loading capacity of the saturated MDEA solution, after four hours, was determined to be 0.979 mol CO_2/mol amine. This value can be used as a benchmark to compare the results obtained in both pilot units. The results presented here, as well as the ones for the other scales described thereafter, are the averaged values from a series of three tests each.

Results from Pilot Unit Scale trials

For the smaller pilot unit, using the same dosing method, the obtained results are shown in Table 4 and represented graphically in Figure 8.

Table 4. Results from CO_2 absorption by the aqueous methyldiethanolamine (MDEA) solution at 20% (w/w) in the small pilot unit

Absorption Time (min)	Loading Capacity (α) (mol CO_2/mol amine)
30	0.529
60	0.676
90	0.824
120	0.824
150	0.971
180	1.029
210	1.089
240	1.118
270	1.118
300	1.118

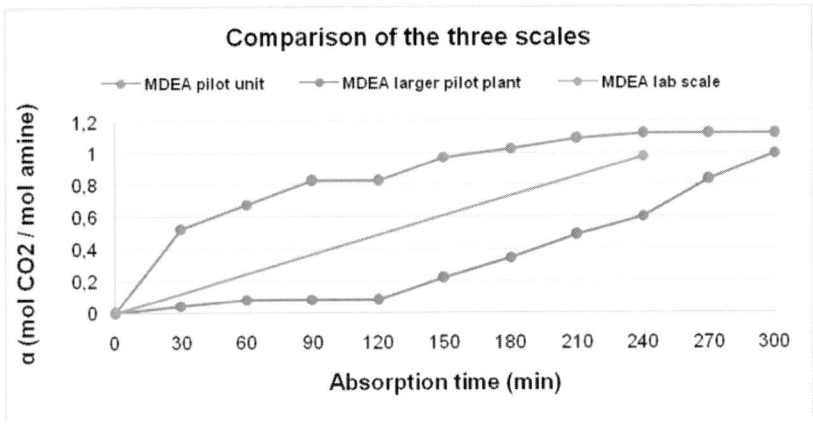

Figure 8. Comparison of the obtained results for CO_2 loading capacity (α) over the absorption time for the tested aqueous MDEA solution (20 wt%) in the three different testing scales.

In the tests performed in the larger pilot plant, the precipitation method was also used to determine the CO_2 loading capacity by the aqueous MDEA solutions in order to assess whether the results obtained in this pilot unit would be in accordance with the ones obtained in the small pilot unit. The obtained results are shown in the Table 5 and graphically in Figure 8.

Table 5. Results from CO_2 absorption by the aqueous MDEA solution at 20% (w/w) by the precipitation method in the larger pilot unit

Absorption Time (min)	Loading Capacity (α) (molCO$_2$/molamine)
30	0.046
60	0.084
90	0.082
120	0.079
150	0.218
180	0.347
210	0.486
240	0.602
270	0.831
300	0.998

Discussion on the Laboratory Scale Trials

It should be noted that MDEA is a tertiary amine and does not react directly with CO_2, but acts as a base, catalyzing the hydration of CO_2, which means that its equilibrium CO_2 loading capacity is approximately 1.0 mol CO_2/mol amine. Therefore, the results obtained at laboratory scale (α = 0.979 mol CO_2/mol amine) are consistent with what was to be expected [24].

Discussion on the Pilot Unit Scale Trials

When analyzing the obtained results, it can be concluded that, by using the smaller pilot unit, it was possible to achieve a CO_2 absorption profile nearly in accordance with the theoretical basis, whereas the MDEA aqueous solution reached CO_2 saturation around the 240th min; although the actual obtained value was slightly higher than it was expected (1.0 mol CO_2/mol amine), this could be attributed to the experimental inaccuracy of the precipitation method used to determine CO_2. Comparing the obtained results for the three scales, which are graphically shown in Figure 6, marked differences appear between the absorption patterns regarding the tests performed in each pilot unit. It should be noted that, although the loading capacity is still increasing, in the larger pilot unit, from the 270th to the 300th min, it seems to be enough, for comparison purposes, to use the loading capacity at 300 min of operation.

These differences may result from process scale-up. In fact, scaling up any process can result in several difficulties for chemical engineers due to the fact that the relative importance of process variables affecting performance considerably increases. In fact, process variables, such as CO_2 injection point in the absorption column, reactant flow, contact time between the reactants throughout the absorption column, temperature and pressure control, and maintenance, have a marked influence on the performance of the units. These variables are somewhat easier to control at laboratory scale and at a small pilot unit, but its significance becomes considerably relevant for larger scales, as noticed by Zlokarnik [25]. This can be easily concluded by comparing the results from both pilot plant units presented here. Although theoretically at the same pressure and temperature (two variables which were kept constant for all performed tests), the saturated CO_2 loading should be the same, regardless of the scale used. However, this does not happen in practice, as demonstrated by this study. In this study, other variables such as column internal packing, pressure drops, residence time, and flooding conditions were kept approximately constant, in spite of scale differences. Nevertheless, it should be noted that, for this preliminary analysis, mass transfer coefficients have not been calculated.

As stated previously, the small pilot unit has a column height of 15 dm, while the larger pilot unit has a column height of 110 dm, representing an increase of 7.3 times, and the inner column diameter increased from 0.505 dm to 0.750 dm in the larger pilot unit, representing an increase of 1.5 times. This results in quite different volumes for the columns of 3 L in the small pilot unit and 62 L in the larger pilot unit, which represents a scale increase of 20.7. In such a larger volume, it is much more difficult to obtain a good mixing contact between the gas phase (CO_2) and the liquid phase (aqueous amine solution); therefore, the system CO_2 loading capacity (α) will be much slower to reach equilibrium, as observed in this study, as shown in Figure 8. Furthermore, the maximum obtained for the latter situation is even lower than the one obtained in the small pilot plant.

These scale-up effects are, in fact, quite difficult to quantify but tend to considerably affect the performance of the absorption process, particularly when progressing for even larger scales, such as industrial plants, as pointed out by previous authors [25].

However, a way to decrease the significance of the inconvenient scale-up effects will be to complement the scaling-up process by additionally including the use of simulation tools in order to help in the selection of the best operating conditions for the process, contributing to high production yields and thus a more profitable operation, as required for the integration of CCS within current production processes [26]. In fact, the use of steady-state or dynamic models for simulation, taking into account precise geometry factors of the columns, as well as operating factors such as the flow rate of each phase, could help to reduce the inaccuracy of the estimation process, as suggested by Panahi and Skogestad [27]. Particular care should be taken in the scale-up of chemical processes involving unit operations such as absorption and stripping, which are, in fact, affected by a multitude of operational factors, apart from dimensional ones.

NEW PROCESS FOR SIMULTANEOUS REMOVAL OF CO_2, SO_X AND NO_X

The present work describes a process for simultaneous removal of CO_2, SO_X and NO_X from industrial flue gases. The process is related with the conventional CO_2 removal process, using chemical absorption, making also possible the simultaneous removal of SO_X and NO_X through the injection of ozone as oxidizing agent and introducing specific sequestrants.

The conventional CO_2 removal process comprises the following steps:

(a) Contact of the flue gas and the oxidizing agent (ozone), with the solvent used in the absorption of carbon dioxide and with the sequestrants used in the removal of the sulfur and nitrogen oxides, counter currently, inside an absorber;
(b) Regeneration of the solution, which retained the carbon dioxide, by heating inside a regeneration column. A regenerated solution and a stream containing mostly carbon dioxide is obtained.

According to the developed work, the regenerated solution obtained in step b) is reused in step a) as solution for the absorption of carbon dioxide.

Figure 9 presents the scheme of the conventional CO_2 removal process (black line) in comparison with the developed process (red and black process).

Compared to the conventional process for chemical absorption of CO_2, this process also introduces some new steps, as follows:

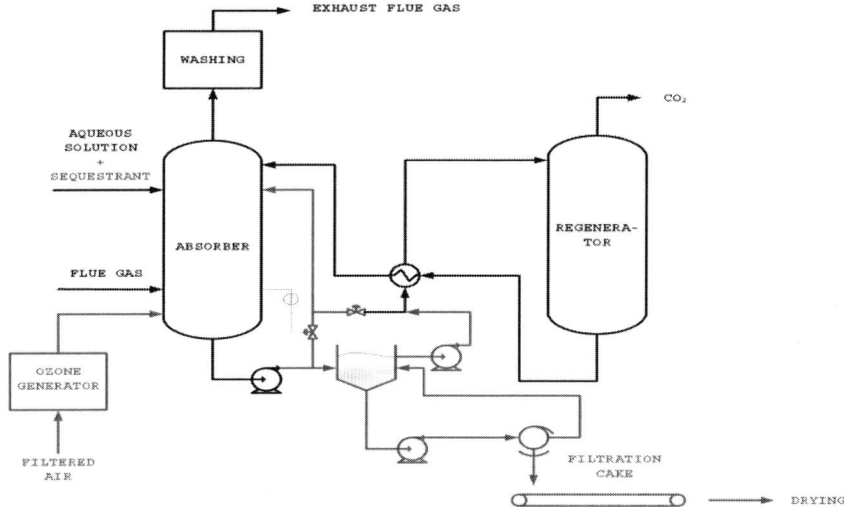

Figure 9. New process scheme [28].

(a₁) Injection of ozone as oxidizing agent and the introduction of specific sequestrants used in the removal of sulfur and nitrogen oxides, in countercurrent, inside the absorption column;

(a₂) Filtration of the precipitated salts resulting from the contact between the sequestrants and the sulfur and nitrogen oxides, after thickening in a clarifier. This step must take place after step (a) of the conventional CO_2 chemical absorption process;

(a₃) Drying of the filtration cake obtained in (a₂), which should take place before stage (b) of the conventional CO_2 chemical absorption process.

According to this new process, the clear phase of step (a₂) can be conducted to step (b). It is important to note that, the agitation caused by gas bubbles, instead of causing dispersion of the slurry, causes a good separation and thickening of the slurry, so it will facilitate the respective separation by filtration.

The solvent used may include, for example, amines, alkanolamines or urea. The sequestering agents may be formed by alkaline compounds, selected from the following: alkali metal salts, alkaline earth metal salts, urea or ammonia.

Apparently this simultaneous removal seemed not to be possible, as the presence of CO_2 could lead to high consumption of metal hydroxide, at a significant loss, as it is likely to form the respective carbonate which precipitates and is subsequently removed in the filter, however carbonate formation is retarded (slower kinetics) by formation of a soluble intermediate and, thus, the kinetic control allows the preferential formation and precipitation of the nitrate, and sulfite occurs before the formation and precipitation of the carbonate [28].

Thus, much of the metal hydroxide is effective in removing NO_X and SO_X, not being destroyed by CO_2, as would be expected. In fact, much of the CO_2 is immediately involved in the complex with the amine due to nucleophilic attack of the amino group to the carbonyl group and so the free CO_2 effective concentration is on average very low, inside the absorber, which, moreover, is important for ensuring effective step absorption.

Description of the Process

Other characteristics and advantages of the new process will be clarified in the following description with reference to Figure 10, showing schematically the method presented in this paper; and Figure 11, which shows schematically and alternative to the new process, where the nitrogen and sulfur oxides removal stage is performed downstream of the conventional CO_2 capture process and, Figure 12, which shows schematically another alternative to the new process, where nitrogen and sulfur oxides removal stage is performed upstream of the conventional CO_2 capture process. Comparing the new developed process with the presented alternatives, it is possible to conclude that the presented process is more viable economically, because it has the advantage of managing the integration of the NO_X and SO_X removal, apart from the conventional CO_2 capture process, without the need of adding any other removal unit, thus reducing the investment and operating costs.

As shown in Figure 10, the flue gas to be treated enters the absorber (C). The flue gas can be produced, for example, from the combustion of hydrocarbons in a boiler or by combustion in a gas turbine. This flue gas contains around 50 to 80% of nitrogen, 5 to 20% of carbon dioxide, 2 to 10% of oxygen and the remnant corresponds to SO_X, NO_X and particles. The flue gas circulates with a pressure between 0.1 MPa abs. and 10 MPa abs. and with a temperature of around 40 °C to 400 °C.

Inside the absorber (C), the flue gas contacts with ozone, produced in the ozone generator (A), and circulates in co-current, with the aqueous solution containing the solvent and the sequestrant counter-current. Ozone is used as oxidizing agent and it is produced from filtered air that goes through the ozone generator (A).

Selection of solvent depends on its carbon dioxide loading capacity, comprising an aqueous solution containing one or more organic compounds capable of absorbing carbon dioxide.

The organic compounds referred above can be, for example, amines (primary, secondary, tertiary, cyclical or not, aromatic or aliphatic), alkanolamines (primary, secondary, tertiary) or urea. The sequestrants are, generally, a saturated aqueous solution of alkaline compounds. These compounds can be, for example, alkali metals salts, alkaline earth metals salts, urea or ammonia. In Figure 10, it is possible to observe to processes: the full line indicates the conventional process and the dashed line indicates the process that allows the simultaneous removal of CO_2, SO_X and NO_X.

The absorber (C) can be a conventional absorption column, as a plate column or a packed bed column.

Inside the absorber (C), the solvent retains and absorbs the carbon dioxide present in the flue gas. NO and SO_2, present in the flue gas are oxidized to NO_2 and SO_3, respectively, through the injection of the oxidizing agent. This oxidation allows better contact with the sequestering agents and results in an easier precipitation of nitrates and sulfates. The cleaned flue gas exits at the top of the absorber, passing through a gas washer (D), and is then released to the atmosphere.

The solvent that retains the carbon dioxide, as well as the nitrates and sulfates from the reaction of NO_X and SO_X with the sequestering agent, exits at the bottom of the absorber (C), feeding the pump (B1).

Afterwards, the mixture can be fed to the decanter (G), allowing the thickening of the solid phase, or, in case of a small or inexistent amount of solids, this mixture can be recycled back to the absorber (C), or, be fed to the heat exchanger (E).

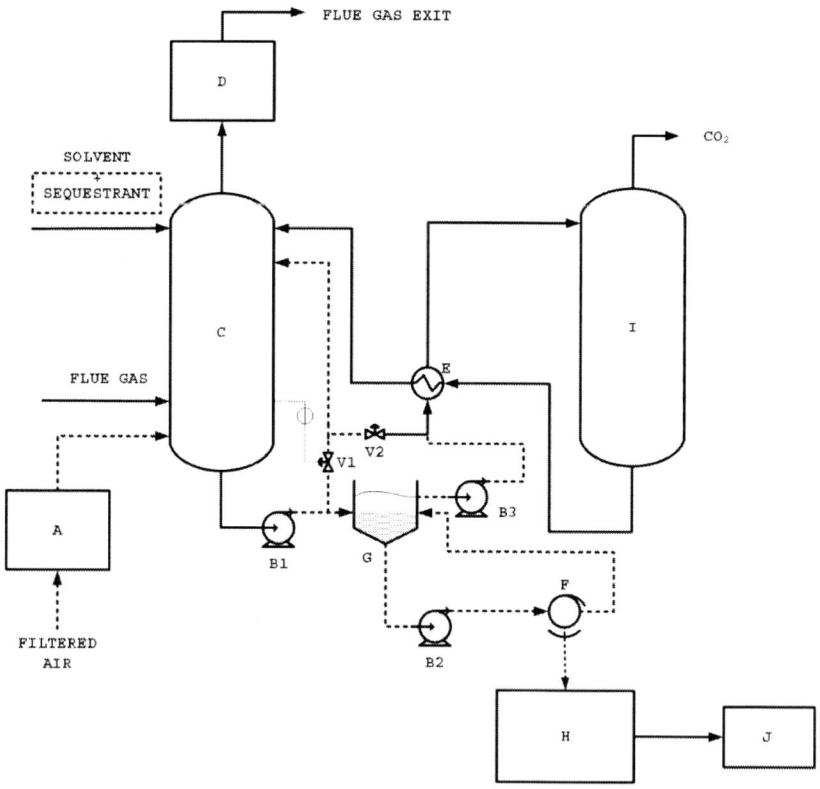

Figure 10. New process scheme [28].

Inside the decanter (G), the separation of the two phases, through settling and thickening of the solid phase takes place. Then, the mixture goes to pump (B2), which conducts the mixture to a rotary vacuum drum filter (F), where the filtration of the solid phase takes place. In the rotary vacuum drum filter (F), the separation of the two phases, allows to obtain a filtration cake, which was sent, by a conveyor (H), to a dryer (J), in order to proceed to its drying process. The liquid phase returns to the decanter (G). The supernatant phase is taken out by the pump (B3).

As mentioned before, if the amount of solids is too low, through the opening of the valve (V1), it is possible to recycle the mixture back to the absorber (C), or, in case of an inexistent solid phase, to flow through the heat exchanger (E), by opening the valve (V2). This line goes together with the current coming from decanter (G), entering the heat exchanger (E).

The heat exchanger (E) allows the heat transfer between the current entering the regenerator (I), and the current that exits the bottom of the regenerator (I) with a higher temperature. This equipment allows the energy integration of the process, because it warms the current that enters the regenerator (I), entering with a higher temperature, decreasing the needed amount of heat to be supplied in order to proceed with the regeneration of the solution, and cools down the exit current, allowing the regenerated solvent to enter the absorber (C) at a temperature closer to the desired temperature for the carbon dioxide absorption, resulting in a more efficient process.

Inside the regenerator (I) the regeneration of the solvent, containing carbon dioxide, takes place. This regeneration is achieved by heating the solvent, and a reaction that is the opposite of the absorption reaction occurs, releasing the absorbed carbon dioxide, and thus regenerating the solvent. The solvent is heated in the bottom of the regenerator (I), where the reboiler is placed. The regeneration temperature will depend on the organic compound in the solvent.

The regenerator (I) can be a column such as, for example, a plate column, thus allowing a better contact between the liquid phase and gaseous phase, resulting in the increase of the solvent regeneration efficiency.

Then, the regenerated solvent flows through the heat exchanger (E), in order to cool down by changing heat with the current that is entering the regenerator (I), and it is recycled to the absorber. From the top of the regenerator (I), a gaseous current that contains predominantly carbon dioxide, exits and goes for ulterior treatment.

Figure 11 shows, schematically, an alternative to the main process for simultaneous removal, where the removal of the nitrogen and sulfur oxides is performed downstream of the conventional process for carbon dioxide capture. Flue gas enters the absorber (C), and the flue gas can be produced, for example, from the combustion of hydrocarbons in a boiler or by combustion in a gas turbine. This flue gas contains around 50 to 80% of nitrogen, 5 to 20% of carbon dioxide, 2 to 10% of oxygen and the remaining percentage corresponds to SO_X, NO_X and particles. The flue gas will circulate with a pressure between 0.1 MPa abs. and 10 MPa abs. and with a temperature of around 40 °C to 400 °C.

Inside the absorber (C), the flue gas contacts with the solvent, flowing in counter-current. Selection of solvent depends on its carbon dioxide loading capacity, comprising an aqueous solution containing one or more organic compounds capable of absorbing carbon dioxide. The carbon dioxide removal in Figure 11 is performed in the conventional CO_2 capture and separation process.

The absorber (C) can be a conventional absorption column, as a plate column or a packed bed column.

CO_2 free flue gas exits the absorber (C) at the top, entering the scrubber (S), where, by contacting with the sequestering agents flowing in counter-current along with the oxidizing agent, the removal of nitrogen and sulfur oxides takes place. NO and SO_2, present in the flue gas are oxidized to NO_2 and SO_3, respectively, through the injection of the oxidizing agent. This oxidation allows easier contact with the sequestering agents and results in an easier precipitation of nitrates and sulfates.

The organic compounds referred above can be, for example, amines (primary, secondary, tertiary, cyclical or not, aromatic or aliphatic), alkanolamines (primary, secondary, tertiary) or urea. The sequestrants are, generally, saturated aqueous solutions with alkaline compounds. These compounds can be, for example, alkali metals salts, alkaline earth metals salts, urea or ammonia. Figure 11 shows the overall process: the full line indicating conventional process and the dashed line, indicating the process that allows the simultaneous removal of CO_2, SO_X and NO_X.

Nitrates and sulfates, from the reaction of NO_X and SO_X with the sequestering agent, exits at the bottom of the absorber (S), feeding the pump (B1). Afterwards, the mixture can be fed to the decanter (G), allowing the thickening of the solid phase.

Inside the decanter (G), the separation of the two phases, through settling and thickening of the solid phase, takes place. Then, the mixture goes to pump (B2), which conducts the mixture to a rotary vacuum drum filter (F), where the filtration of the solid phase takes place. In the rotary vacuum drum filter (F), the separation of the two phases, allowed to obtain a filtration cake, that was sent, by a conveyor (H), to a dryer (J), in order to proceed to its drying process. The liquid phase returns to the decanter (G). The supernatant phase is taken out by the pump (B3) and recycled back to the scrubber (S).

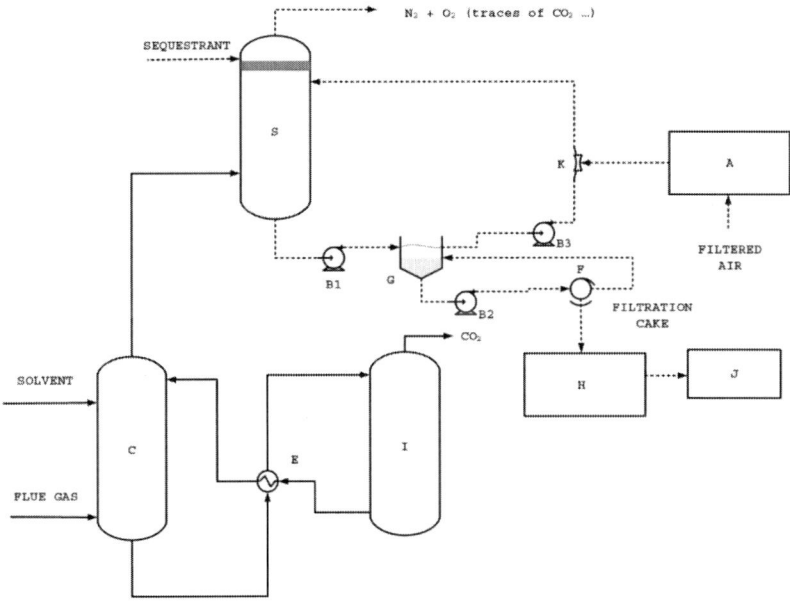

Figure 11. Alternative scheme for the new process [28].

The oxidizing agent injected in the scrubber (S) is produced from filtered air that goes through the ozone generator (A). The produced ozone flows inside the scrubber (S), through a Venturi tube (K).

Figure 12 shows, schematically, an alternative to the main process for simultaneous removal, where the removal of the nitrogen and sulfur oxides is performed upstream of the conventional process for carbon dioxide capture. Flue gas enters the scrubber (S) and can be result, for example, from the combustion of hydrocarbons in a boiler or by combustion in a gas turbine. This flue gas contains around 50 to 80% of nitrogen, 5 to 20% of carbon dioxide, 2 to 10% of oxygen and the remaining percentage corresponds to SO_X, NO_X and particles. The flue gas will circulate with a pressure between 0.1 MPa abs. and 10 MPa abs. and with a temperature of around 40 °C to 400 °C.

Flue gas enters the scrubber (S), where, by contacting with the sequestering agents that flow in counter-current, along with the oxidizing agent, the nitrogen and sulfur oxides removal takes place. NO and SO_2, present in the flue gas, are oxidized to NO_2 and SO_3, respectively, through the injection of the oxidizing agent.

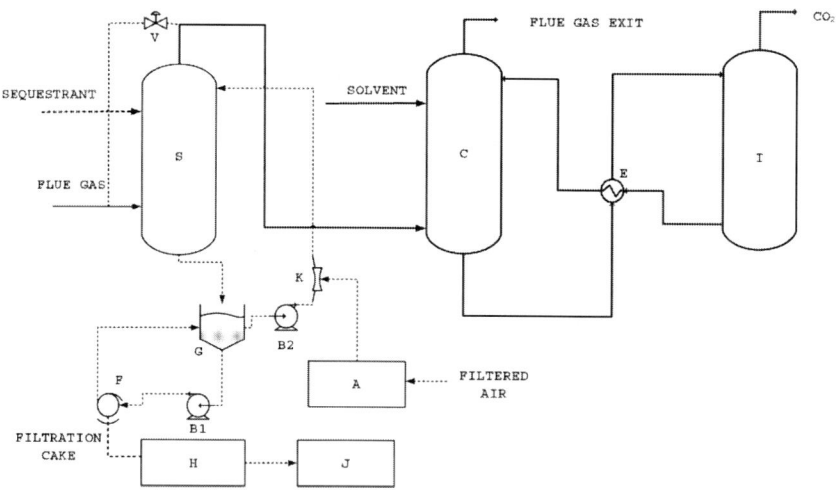

Figure 12. Alternative scheme for the new process [28].

This oxidation allows the contact with the sequestering agents and results in an easier precipitation of nitrates and sulfates. Figure 12 shows the whole process: the full line, conventional process and the dashed line, the process that allows the simultaneous removal of CO_2, SO_X and NO_X.

The sequestrants are, generally, saturated aqueous solutions of alkaline compounds. These compounds can be, for example, alkali metals salts, alkaline earth metals salts, urea or ammonia.

The nitrates and sulfates from the reaction of NO_X and SO_X with the sequestering agent, exits at the bottom of the absorber (S), feeding the decanter (G), thus allowing the thickening of the solid phase.

Inside the decanter (G), the separation of the two phases, through settling and thickening of the solid phase takes place. Then, the mixture goes to pump (B1), which conducts the mixture to a rotary vacuum drum filter (F), where the filtration of the solid phase takes place. In the rotary vacuum drum filter (F), the separation of the two phases, allowed to obtain a filtration cake, that was sent, by a conveyor (H), to a dryer (J), in order to proceed to its drying process. The liquid phase returns to the decanter (G). The supernatant phase is taken out by the pump (B2) and recycled back to the scrubber (S).

The oxidizing agent injected in the scrubber (S) is produced from filtered air that goes through the ozone generator (A). The produced ozone flows inside the scrubber (S), through a Venturi tube (K).

Cleaned flue gas exits the scrubber (D) at the top, entering the absorber (C), where the flue gas contacts with the solvent that flow in counter-current. Selection of solvent depends on its carbon dioxide loading capacity, comprising an aqueous solution containing one or more organic compounds capable of absorbing carbon dioxide. The carbon dioxide removal shown in Figure 12 takes place as in the conventional CO_2 capture and separation process.

The organic compounds referred above can be, for example, amines (primary, secondary, tertiary, cyclical or not, aromatic or aliphatic), alkanolamines (primary, secondary, tertiary) or urea.

The following test was performed in order to prove this concept on a pilot-unit. The prepared solvent consists on an aqueous solution containing 30 percent, in weight, of MEA (monoethanolamine). The used sequestrant was an aqueous "slurry" containing 10 percent, in weight, of sodium hydroxide. The flue gas to be treated circulated at a flow of 1.2 Nm^3/h and having the following composition: 14.75% CO_2, 76.23% N_2, 5.81% O_2, 480 ppm SO_X, 50 ppm CO and 420 ppm NO_X (as NO_2). In order to remove CO_2, NO_X and SO_X, ozone (as oxidizing agent) was diluted in oxygen and introduced, at a mass percentage of 5%, along with the solvent, inside the absorber (C), at a flow of 0.04 m^3/h. The filtration of the solids formed inside the absorber (C), originated from the reaction of the sequestrant with NO_X and SO_X, takes place on a rotary vacuum drum filter (F), with a vacuum pressure of 0.04 MPa. Regeneration of the solvent, coming from the bottom of the absorber (C), takes place in the regenerator (I) by distillation. The temperatures inside the regenerator (I), at the top and at the bottom, are 120 °C and 40 °C, respectively. The amount of obtained solids ranges from 0.4 kg to 0.6 kg.

Companies using CO_2 and/or acid gases absorption equipment, actively look for NO_X and SO_X removal systems. Unlike current $DeNO_X$ systems that are very expensive in terms of investment and operative costs, this new process presents a more moderate investment cost, as well as lower operating costs, thus resulting in very significant savings.

REFERENCES

[1] Yang, H; Xu, Z; Maohong, F; Gupta, R; Slimane, R; Bland, A; Wright, I. Progress in carbon dioxide separation and capture: a review. *J. Env. Sci.*, 2008, 20, 14-27.

[2] Lin, S; Shreyer, C. Carbon dioxide absorption by amines: system performance predictions and regeneration of exhausted amine solution. *Env. Tech.*, 2000, 21, 1245-1254.

[3] Figueroa, J; Fout, T; Plasynski, S; McIlvried, H; Srivastava, R. Advances in CO_2 capture technology – The US Department of Energy's Carbon Sequestration Program. *Int. J. Greenhouse Gas Control.*, 2008, 2, 9-20.

[4] Bachelor, T; Tsochinda, P. *Development of low-cost amine-enriched solid sorbent for CO_2 capture.*, 2012, 33, 2645-2651.

[5] Gomes, J. *Carbon Dioxide Sequestration Technologies.* New York: Nova Publishers Inc. Eds., 2013.

[6] Johnsen, K; Helle, K; Myhrvold, T. Scale-up of CO_2 capture processes: The role of technology qualification. *Energy Proc.*, 2009, 1, 163-170.

[7] Wall, T. Combustion processes for carbon capture, *Proc. Comb. Inst.*, 2007, 31, 31-47.

[8] Puxty, G; Rowland, R; Attalla, M. Comparison of the rate of CO_2 absorption into aqueous ammonia and monoethanolamine. *Chem. Eng. Sci.*, 2010, 65, 915-922.

[9] Greer, T; Bedelbayev, A; Igreja, J; Gomes, J; Lie, B. A simulation study on the abatement of CO_2 emissions by de-absorption with monoethanolamine. *Env. Tech.*, 2010, 31, 107-115.

[10] Paul, S; Ghoshal, A; Mandal, B. Kinetics of absorption of carbon dioxide into aqueous blends of 2-(1-Piperazinyl)ethylamine and N-methyldiethanolamine. *Chem. Eng. Sci.*, 2009, 64, 313-321.

[11] Gray, M; Soong, Y; Champagne, K; Pennline, H; Baltrus, J; Stevens, R; Khatri, R; Chuang, S; Filburn, T. CO_2 capture by amine-enriched fly ash carbon sorbents. *Fuel Proc. Tech.*, 2008, 86, 1449-1455.

[12] Chakraborty, A; Astarita, G; Bischoff, K. CO_2 absorption in aqueous solutions of hindered amines. *Chem. Eng. Sci.*, 1986, 41, 997-1000.

[13] Singh, P; Nieder, J; Versteeg, G. Structure and activity relationships for amine based CO_2 absorbents – I. *Int. J. Greenhouse Gas Control.*, 2007, 1, 5-10.

[14] Singh, P; Nieder, J; Versteeg, G. Structure and activity relationships for amine based CO_2 absorbents – II. *Chem. Eng. Res. & Dev.*, 2009, 87, 135-144.

[15] Santos, S; Gomes, J; Bordado, J. *Comparative Study of Amine Solutions used in Absorption/Desorption Cycles of CO_2*. Saarbrucken: LAP Lambert Academic Publishing, 2014.

[16] Santos, A. *Estudos de Absorção de CO_2 em Soluções Aminadas* [MSc Thesis in Chemical Engineering]. Lisboa, Portugal: Instituto Superior de Engenharia de Lisboa; Lisboa, 2012, (in Portuguese). [*CO2 Absorption Studies in Amined Solutions*]

[17] Li, M; Chang, B. Solubilities of Carbon Dioxide in Water + Monoethanolamine + 2-Amino-2-methyl-1-propanol. *J. Chem. Eng. Data.*, 1994, 39, 448-452.

[18] Rao, A; Rubin, E. A technical, economic and environmental assessment of amine-based CO_2 capture technology for power plant greenhouse gas control. *Env. Sci. Tech.*, 2002, 36, 4457-4475.

[19] MacDowwell, N; Florin, N; Buchard, A; Hallet, J; Galindo, A; Adjiman, C; Williams, C; Shah, N; Fennell, P. An overview of CO_2 capture technologies. *Energy Env. Sci.*, 2010, 3, 1645-1669.

[20] Supap, T; Idem, R; Tontiwachwuthikul, P; Saiwan, C. Kinetics of sulfur dioxide- and oxygen-induced degradation of aqueous monoethanolamine solution during CO_2 absorption from power plant flue gas streams, *Int. J. Greenhouse Gas Control.*, 2009, 3, 133-142.

[21] Li, M; Chang, B. Solubilities of carbon dioxide in Water + Monoethanolamine + 2-Amino-2-methyl-1-propanol. *J. Chem. Eng. Data*, 1994, 39, 448–452.

[22] Santos, A. *Estudos de Absorção de CO_2 em Soluções Aminadas*. Master's Thesis, Instituto Superior de Engenharia de Lisboa, Lisboa, Portugal, 2012, (In Portuguese). [*CO2 Absorption Studies in Amined Solutions*]

[23] Santos, S; Gomes, J; Bordado, J. *Comparative Study of Amine Solutions Used in Absorption/Desorption Cycles of CO_2*; LAP Lambert Academic Publishing: Saarbrucken, Germany, 2014.

[24] Budzianowski, W. Single solvents, solvent blends and advanced solvent systems in CO_2 capture by absorption: A review. *Int. J. Glob. Warm.*, 2015, 7, 184–225.

[25] Zlokarnik, M. *Scale up in Chemical Engineering*, 2nd ed.; Wiley-VCH: Weinheim, Germany, 2006.

[26] Budzianowski, W. Low-carbon power generation cycles: The feasibility of CO_2 capture and opportunities for integration. *J. Power Technol.*, 2011, 91, 6–13.

[27] Panahi, M; Skogestad, S. Economically efficient operation of CO_2 capturing process part I: Self-optimizing procedure for selecting the best controlled variables. *Chem. Eng. Process.*, 2011, 50, 247–253.

[28] Santos, SP; Duarte, AP; Bordado, JC. Processo de Remoção Simultânea *In Situ* de CO_2, NO_X e SO_X, Patente Provisória PT 108544, 2015, (in Portuguese).

BIOGRAPHICAL SKETCHES

João Gomes

Affiliation: ISEL – Instituto Superior de Engenharia de Lisboa and CERENA – Centro de Ambiente e Recursos Naturais, IST-Instituto Superior Técnico, Lisboa, Portugal

Education: BSc Chemical Eng.; PhD Chemical Eng.; Habilitation Chemical Eng.

Research and Professional Experience: Previously teaching in: IST-Lisbon University, ISCTE-Lisbon University, ULHT-Universidade Lusófona and UATLA-Universidade Atlântica. Researcher of LNETI – National Laboratories for Energy, Portugal. Researcher and Deputy Director of ISQ – Instituto de Soldadura e Qualidade.

Professional Appointments: Professor of ISEL; Senior Researcher of CERENA.

Publications from the Last 3 Years:

Catarino, M; Ferreira, E; Dias, A; Gomes, JFP. "Dry washing biodiesel purification using fumed silica sorbent", *Chemical Engineering Journal*, 123930, (2019). DOI: 10.106/j.cej.2019.123930.
Catarino, M; Martins, S; Dias, A; Pereira, M; Gomes, JFP. "Calcium diglyceroxide as a catalyst for biodiesel production", *Journal of Environmental Chemical Engineering*, 7(3), 103099, (2019). DOI: 10.1016/j.jece.2019.103099.

Catarino, M; Ramos, M; Dias, A; Santos, T; Puna, J; Gomes, JFP. "Calcium rich food wastes based catalysts for biodiesel production", *Waste and Biomass Valorization*, 8, 1699-1707, (2017).
DOI: 10.1007/s12649-017-9988-8.

Dias, A; Puna, J; Correia, M; Gomes, JFP; Bordado, J. "Strontium-doped lime catalysts for biodiesel production. Activity and stability during soybean oil methanolysis", *Current Topics in Catalysis*, 13, 35-42, (2017).

Dias, A; Ramos, M; Puna, J; Gomes, JFP; Bordado, J. "Review on biodiesel production processes and sustainable raw materials", *Energies*, 12, 4408, (2019).
DOI: 10.3390/en12234408.

Ferreira, O; Rijo, P; Gomes, JFP; Santos, R; Monteiro, S; Vilas-Boas, C; Correia-da-Silva, M; Almada, S; Alves, L; Bordado, J; Silva, E. "Biofouling inhibition with grafted econea biocide: toward a non releasing eco-friendly multiresistant antifouling coating", *ACS Sustainable Chemistry & Engineering*, (2019). DOI: 10.1021/acssuschemeng.9b04550.

Gomes, JFP; Miranda, R. "Determination of "safe" and "critical" nanoparticles exposure to welders in a workshop", *Journal of Toxicology and Environmental Health – A*, 80(13-15), 767-775, (2017). DOI:10.1080/15287394.2017.1286904.

Gomes, JFP; Miranda, R; Oliveira, J; Esteves, H; Albuquerque, P. "Evaluation of the amount of nanoparticles in LASER additive manufacture/welding", *Inhalation Toxicology*, 31, (2019). DOI: 10.1080/08958378.2019.1621965.

Gomes, JFP; Miranda, R; Porro, J; Ocaña, J. "Experimental characterization of nanoparticles emissions during Laser Shock Processing of AA6061, AISI304 and Ti6Al4V", *Revista Metalurgia Madrid*, 53(4), (2017). DOI: 10.3989/revmetalm.104.

Gonçalves, A; Puna, J; Guerra, L; Rodrigues, J; Gomes, JFP; Santos, M; Alves, D. "Towards the Development of Syngas/Biomethane Electrolytic Production, Using Liquefied Biomass and Heterogeneous Catalyst", *Energies*, 12, 3787, (2019). DOI: 10.3390/en12193787.

Guerra, L; Moura, K; Rodrigues, J; Gomes, JFP; Puna, J; Santos, T., "Synthesis gas production from water electrolysis, using the Electrocracking concept", *Journal of Environmental Chemical Engineering*, 6, 604-609, (2018). DOI: 10.1016/j.jece.2017.11.033.

Guerra, L; Rossi, S; Rodrigues, J; Gomes, JFP; Puna, J; Santos, M. "Methane production by a combined Sabatier reaction/water electrolysis process", *Journal of Environmental Chemical Engineering*, 6, 671-676, (2018). DOI: 10.1016/j.jece.2017.12.066.

Ozkan, S; Puna, J; Gomes, JFP; Cabrita, T; Palmeira, V; Santos, M. "Preliminary study on the use of biodiesel obtained from waste vegetable oils for blending with hydrotreated kerosene fossil fuels using calcium oxide from natural waste materials as heterogeneous catalyst", *Energies*, 12, 4306, (2019). DOI: 10.3390/en12224306.

Pacheco, R; Gomes, JFP; Miranda, R; Quintino, L. "Evaluation of the amount of nanoparticles emitted in welding fume from stainless steel using different shielding gases", *Inhalation Toxicology*, 29(6), 282-296, (2017). DOI: 10.1080/08958378.2017.1358778.

Ramos, M; Catarino, M; Puna, J; Gomes, JFP. "Solvent Assisted Biodiesel Production by Co-processing Beef Tallow and Soybean Oil over Calcium Catalysts", *Waste and Biomass Valorization*, (aceite, Nov. 2019). DOI: 10.1007/s12649-019-00903-7.

Santos, S; Nobre, L; Gomes, JFP; Puna, J; Quinta-Ferreira, R; Bordado, J. "Soybean Oil Transesterification for Biodiesel Production with Micro-Structured Calcium Oxide (CaO) from Natural Waste Materials as a Heterogeneous Catalyst", *Energies*, 12, 4670, (2019). DOI: 10.3390/en12244670.

Santos, T; Gomes, JFP; Puna, J. "Liquid-liquid equilibrium for ternary systems containing biodiesel, metanol and water", *Journal of Environmental Chemical Engineering*, 6, 984-990, (2018). DOI: 10.1016/j.jece.2017.12.068.

Samuel Santos

Affiliation: CERENA – Centro de Ambiente e Recursos Naturais, IST-Instituto Superior Técnico, Lisboa, Portugal

Education: BSc Chemical Eng.; MSc Chemical Eng.

Research and Professional Experience: PhD student in Chem. Eng.

Publications from the Last 3 Years:

Santos, S; Nobre, L; Gomes, JFP; Puna, J; Quinta-Ferreira, R; Bordado, J. "Soybean Oil Transesterification for Biodiesel Production with Micro-Structured Calcium Oxide (CaO) from Natural Waste Materials as a Heterogeneous Catalyst", *Energies*, 12, 4670, (2019). DOI: 10.3390/en12244670.

João Bordado

Affiliation: CERENA – Centro de Ambiente e Recursos Naturais, IST-Instituto Superior Técnico, Lisboa, Portugal

Education: BSc Chemical Eng.; MSc Chemical Eng.; PhD Chemical Eng.; Habilitaion Chemical Eng.

Research and Professional Experience: teaching in IST; Process Engineer at Quimigal; Head of R&D Hoechst; Full professor at IST

Publications from the Last 3 Years:

Dias, A; Puna, J; Correia, M; Gomes, JFP; Bordado, J. "Strontium-doped lime catalysts for biodiesel production. Activity and stability during soybean oil methanolysis", *Current Topics in Catalysis*, 13, 35-42, (2017).

Dias, A; Ramos, M; Puna, J; Gomes, JFP; Bordado, J. "Review on biodiesel production processes and sustainable raw materials", *Energies*, 12, 4408, (2019). DOI: 10.3390/en12234408.

Ferreira, O; Rijo, P; Gomes, JFP; Santos, R; Monteiro, S; Vilas-Boas, C; Correia-da-Silva, M; Almada, S; Alves, L; Bordado, J; Silva, E. "Biofouling inhibition with grafted econea biocide: toward a non releasing eco-friendly multiresistant antifouling coating", *ACS Sustainable Chemistry & Engineering*, (2019). DOI: 10.1021/acssuschemeng.9b04550.

Santos, S; Nobre, L; Gomes, JFP; Puna, J; Quinta-Ferreira, R; Bordado, J. "Soybean Oil Transesterification for Biodiesel Production with Micro-Structured Calcium Oxide (CaO) from Natural Waste Materials as a Heterogeneous Catalyst", *Energies*, 12, 4670, (2019). DOI: 10.3390/en12244670.

Ana Duarte

Affiliation: Universidade Atlântica, Barcarena, Portugal

Education: BSc Mats. Eng.; MSc Mats. Eng.; PhD Mats. Eng.

Research and Professional Experience: research in IST; teaching in Universidade Atlântica

In: Carbon Dioxide Emissions
Editor: Asia Santana
ISBN: 978-1-53617-763-3
© 2020 Nova Science Publishers, Inc.

Chapter 2

SOIL CO_2 EMISSION IN BRAZIL

Gabriel Ribeiro Castellano[*]
Earth Sciences and Exact Sciences Institute (IGCE), UNESP,
Rio Claro, SP, Brazil

ABSTRACT

The signficant increase in atmospheric CO_2 in the last century is primarily due to fossil fuel combustion (36.6 Gt of CO_2 in 2018) and land-use change/deforestation (5.5 Gt of CO_2 per year from 2009 to 2018). In Brazil, agricultural activities account for 22% of total CO_2 emission. Land use change, the main cause of CO_2 emission in the country, accounts for 51%. These changes occur mainly in forests and savannas, because their soil and climate conditions are ideal for high-yield agricultural production. Changes in land cover significantly alter physical, biological, and chemical characteristics of soils. Soil CO_2 emissions (FCO_2) is a result of physical and biochemical processes that determine CO_2 production and transport from soil to atmosphere. CO_2 production is related to microorganism activity and plant root respiration, whereas CO_2 transport is associated to the physical structure of the soil, especially its porosity, which affects soil gas flux. Based on pooled data from FCO_2 research carried out in Brazil from 1990 to 2019 with IRGA (infra-red gas analyzer), this study aims to assess the effects of land use change on soil carbon flux in Brazil, in

---

[*] Corresponding Author's E-mail: grcastellano@gmail.com.

addition to contributing to the body of knowledge about carbon stock balance in tropical and subtropical domains. A bibliographical review was conducted and data from research done in the Amazon Forest, Atlantic Forest, Cerrado (South America savanna), and agricultural crops were pooled. FCO_2 in the Amazon Forest ranged from 3.2 to 6.4 μmol CO_2 m^{-2}s^{-1}; several studies reported a significant linear correlation ($p < 0.05$) between FCO_2 and soil moisture. FCO_2 in the Atlantic Forest ranged from 0.51 to 3.86 μmol CO_2 m^{-2} s^{-1}, indicating a significant linear correlation with soil moisture ($r = 0.55$, $p < 0.0001$). FCO_2 in the Cerrado was 2.55 μmol and 0.86 μmol CO_2 m^{-2} s^{-1} CO_2 m^{-2} s^{-1} in the rainy and dry seasons, respectively. In agricultural crops in Brazil's southeast (São Paulo State), FCO_2 ranged from 1.19 up 5.3 mol CO_2 m^{-2} s^{-1}. Most of these studies were conducted in sugarcane plantations using spatial variability analysis, while those carried out in natural areas focused on temporal variability. Studies in Brazil indicate that soil moisture has the most important temporal influence on FCO_2 in forests and savannas. In agricultural crops, FCO_2 values are affected by soil and agricultural management practices. The implications of the land cover change in Brazil need to be discussed.

Keywords: soil respiration, soil CO_2 emission, Amazon forest, Atlantic forest, agricultural crops

INTRODUCTION

Greenhouse gas (GHG) emission has become one of the top environmental concerns in recent times, with three of the greenhouse gases directly linked to agricultural and forestry activities. Carbon dioxide (CO_2), methane (CH_2), and nitrous oxide (N_20) contribute 49%, 30%, and 21% of total emissions, respectively. In particular, the concentration of CO_2 in the atmosphere has risen from 280 ppm to about 407 ppm since the early days of the industrial revolution (Denman et al. 2007; Kuntoro and Wahyu 2009; Food and Agriculture Organization of the United Nations 2016; Word Meteorological Organization 2019).

One of the main causes of growing CO_2 concentration in the atmosphere is the growth in anthropogenic activity, leading to the replacement of native plant species, the cutting, felling, and burning of trees and vegetation for economic gain, and changes in land cover and use. Change in land use and agriculture combined account for 21% of total CO_2 emission to the atmosphere (Sabine et al. 2004; Food and Agriculture Organization of the United Nations 2016). Estimates show that anthropogenic activity occurs mainly in forests and savannas, because their soil and climate conditions are ideal for high-yield agricultural production.

The increase in atmospheric CO_2 in the last century has been primarily due to emissions from fossil fuel combustion (36.6 Gt of CO_2 in 2018) and deforestation and land-use changes (5.5 Gt of CO_2 per year from 2009 to 2018). In Brazil, deforestation and expansion of agricultural frontiers in the Cerrado (South American savanna) and Amazon regions for agricultural and cattle-raising purposes have contributed about 2 Gt of CO_2-eq per year, thus making the country one of the world's largest CO_2 emitters, after China, the United States, and the European Union (Word Meteorological Organization 2019).

If the predicted climate change materializes, the impact on forests will be long-lasting and profound. The extent to which each ecosystem will be affected may vary from region to region. That is, the composition, distribution, growth, and dynamics of each forest/savanna ecosystem will be differently affected by climate change. In this context, new research on forest restoration has emerged, especially studies related to the quantification of environmental services provided by reforestation with native species, carbon fixation via this process, and its effectiveness in reducing atmospheric CO_2 levels (Intergovernmental Panel Climate Change 2001; Food and Agriculture Organization of the United Nations 2001; Foster and Mello 2007).

Tropical ecosystems account for 20-25% of the world's terrestrial carbon. A significant amount of carbon is stored in the soil (Schlesinger 1997), which has a major impact on the biogeochemical cycles that contribute to the regulation of global warming. For this reason, many studies have been conducted on the influence of this element on soils with a view to contributing data to climate change models. In this context, Kutsch et al. (2010) pose some questions regarding the capacity of ecosystems to sequester CO_2, such as:

1. How much CO_2 can be sequestered by the soil in each of the world's ecosystems? And how long does this carbon stay in the soil?
2. Does a rise in net primary production of an ecosystem—due to mounting atmospheric CO_2 concentration associated to anthropic activities, e.g., nitrogen fertilization—increase litter production and, as a result, carbon stock in soils?

Forest biomes are efficient carbon storers; they account for approximately half of the total carbon fixed by terrestrial vegetation. Boreal forests are responsible for 26% of total terrestrial carbon stocks, while temperate and tropical forests account for 20% and 7%, respectively. Brazil is the fifth largest country in territorial extension, with approximately 5.7% and 47.3% of the world's and South American surface areas, in that order. It also has an impressive natural heritage, which places it at the top of the list of megadiverse countries, i.e., those with the largest number of plant and animal species (Dixon et al. 1994; Campanili and Schaffer 2010).

Among Brazil's main biomes, the Atlantic Forest—which originally covered an area of approximately 1,300,000 km², spanning throughout 17 Brazilian states—has now only 27% of its original coverage. It consists of a set of forest formations, in addition to associated ecosystems (e.g., natural fields and mangroves), with remnants distributed in thousands of vegetation fragments, which still maintain high levels of biodiversity and provide invaluable environmental services for water source protection, slope containment, and climate regulation. Due to spreading throughout several regions of the Brazilian territory, the Atlantic Forest has undergone major economic changes (e.g., dependence on agricultural processes and wood removal from its upper stratum). In light of these changes, it is unwise to assume that even today's remnants have never suffered strong anthropic pressures in the past (Rodrigues 1999; Campanili and Schaffer 2010).

The recovery of the Atlantic Forest, therefore, is relevant, not only with regard to its biodiversity and other related attributes—which have been explicitly addressed in the Atlantic Forest Restoration Pact (Pacto 2019)—, but also because it plays an important role in CO_2 regulation of ecosystems. According to the protocol established in the aforementioned pact, 15 million hectares of its original coverage are to be restored by 2050. This process will bring about changes in land use, which should alter CO_2 balances, regionally and globally. Among the protocol's priorities is the appraisal of the environmental or ecosystem services provided to society by its remaining and restored areas, reinforcing their importance to the quality of life and means of production, taking advantage of opportunities at the carbon and water markets.

However, for these services to be properly appraised, it is necessary to conduct a broad and in-depth study on biogeochemical carbon cycles in the Atlantic Forest, focusing on the evaluation and characterization of FCO_2, which constitutes an important indicator of soil environmental quality in addition to providing guidelines for planting and restoration plans. Rainforests are considered the most productive and most diverse ecosystems on Earth. Among them, the Amazon rainforest accounts for 60% of the world's tropical forest area, equivalent to 5×10^6 km^2 (Dixon et al. 1994).

The vegetation cover of Brazil's Amazon ecosystems is predominantly forest (48.8% dense forest and 27.1% open forest), complemented with smaller swaths of savannas (17.1%) and other natural fields and floodplains (7.0%). It has been estimated that 58 million hectares, corresponding to 16% of the Brazilian Amazon forest, have already been destroyed. In light of the heterogeneity of the Amazon rainforest, with its unique soil and carbon dynamics, it is important to investigate FCO_2 because of soil respiration. According to Houghton, Skole, and Nobre (2000), a typical forest in the Amazon region sustains an average of 360 tons of plant biomass per hectare, corresponding to 170 tons of C per hectare.

Rainforest soils play a significant role in the dynamics of chemical and physical processes of the atmosphere, since they act as source and sink for various trace gases, especially CO_2 (Keller, Kaplan, and Wofsy 1986; Goreau and De Mello 1987). It is estimated that as much as 2/3 of the carbon found in mature rainforests is fixed as humid organic matter (Malhi, Baldocchi, and Jarvis 1999). However, little is known about CO_2 dynamics and residence time in the Amazon forest because it comprises a mosaic of different soils whose CO_2 stocks vary significantly.

Likewise, Brazil's savanna, i.e., Cerrado, has great biological diversity. It is the second largest biome in South America, covering an area of approximately 204 million hectares, about 25% of the entire Brazilian territory. In the last four decades, mainly from 1990 to 2011, as many as one million km^2 have been converted into agricultural areas. Recently, 80 million hectares of Cerrado have been classified under various land use types, about 39.5% of its total area. The most recent national inventory, reporting on historical emissions between 1990 and 2014, indicated that land use changes in the Cerrado biome—mainly caused by deforestation, degradation or conversion of natural vegetation for agricultural use—emitted about 89.6 Gg CO_2-eq, accounting for 34.12% of the country's total emission (Lapola et al. 2013; Redo et al. 2013; Bustamante et al. 2014; Beuchle et al. 2015; Ministério da Ciência, Tecnologia e Inovação 2014).

Over the years, the scientific community has been concerned about the Amazon rainforest and how anthropic activities—such as forest burning and deforestation, burning of fossil fuels, and changes in land use (replacement of vegetation cover with pastures, intensive crops or urban developments)—impact the carbon cycle dynamics that trigger climate and environmental changes. Thus, knowledge of the CO_2 flux from soil to the atmosphere is of great importance to assessing the ecosystem's **carbon dynamics** in order to better conceptualize the biogeochemical carbon cycle balance.

SOIL CO_2 EMISSION AND ENVIRONMENTAL VARIABLES

Soil carbon dioxide emission corresponds to CO_2 produced by respiration of roots and microorganisms present in the soil, which are responsible for aerobic decomposition of organic matter (OM) and are influenced by vegetation type and soil characteristics. The main factors influencing *F*CO_2 may be divided into physical characteristics (e.g., soil moisture, temperature, texture, and structure), chemical characteristics (e.g., phosphorus content, C/N ratio, pH), biological characteristics (e.g., microbial activity), and climatic parameters (e.g., temperature, air humidity, and photosynthetically active radiation).

FCO_2 is the result of physical, chemical, and biological processes that interfere with producing and transporting CO_2 from soil to atmosphere. In particular, CO_2 transport is associated to the physical structure of the soil and the conditions under which soil gas flows. The rate of CO_2 transport from soil to atmosphere is influenced by five factors: (i) CO_2 production rate in soil; ii) temperature gradient; iii) CO_2 concentration at the soil-atmosphere interface; iv) physical properties of the soil; and v) atmospheric pressure fluctuation (Sotta 1998).

In soils without vegetation, CO_2 production is linked to microbial activity and its transport from soil to atmosphere is explained by the diffusion equation, controlled by the CO_2 concentration gradient between soil and atmosphere. FCO_2 varies significantly in space and time due to the heterogeneity of the system and the dynamics of the factors controlling them. Despite its importance, little is known about CO_2 flux from soil to atmosphere vis-à-vis seasonal variation and amount of respiration in different types of land cover (Medina, Klinge, and Jordan 1980; Davidson, Trumbore, and Amundson 2000; Feilg, Melilo, and Cerri 1995; Meir et al. 1996; Le Dantec et al. 1999; La Scala et al. 2000a; Fernandes et al. 2002; Panosso et al. 2008).

Research on the correlation between organic matter content in the soil and FCO_2 shows a positive correlation (La Scala et al. 2000b)—mainly due to substrate supply to microbial activity—as well as a negative one (Fang et al. 1998)—as a probable consequence of limitations to the decomposition process related to adverse soil and climate conditions. Soil moisture and temperature are the factors that most influence FCO_2, with soil moisture not always presenting the same patterns of spatial and temporal variability. In general, soil moisture is negatively correlated with FCO_2 in spatial variability studies and positively correlated in temporal variability studies (Xu and Qi 2001; Epron et al. 2004 and 2006; Kosugi et al. 2007; La Scala et al. 2010).

The influence of soil temperature on FCO_2 should be carefully evaluated, specifically regarding the manner in which soil temperature correlates with soil moisture. Research by Xu and Qi (2001) indicated that soil temperature and humidity explained 70% of the temporal variability of

FCO_2, which may be interpreted in light of the correlation between soil moisture and soil temperature. According to Epron et al. (2004), it is common to observe a positive correlation between soil humidity and soil temperature in tropical regions, where the dry season is often colder than the rainy one. For this reason, the bivariate model using temperature and soil moisture is not as adequate to explain the temporal variation of FCO_2 as is the univariate model with soil moisture alone. According to Carbonell-Bojollo et al. (2013), soil temperature changes do not always influence soil CO_2 flux, which is corroborated by a study by La Scala et al. (2010) on spatial variability of soil temperature and FCO_2. Kosugi et al. (2007) reported that soil moisture is more indicated to estimate soil respiration variation in regions where temperature variation is small.

Aggregate stability and soil texture are factors that have a major impact on carbon stocks and FCO_2, mainly because they reflect pore size. Soil porosity and compaction may enable or hinder gas storage and transport. Macro and microporosity percentages affect plant roots and microbial activity responsible for soil respiration (Siqueira Neto et al. 2011; Goutal et al. 2012; Carbonell-Bojollo et al. 2012). A study by Davidson and Swank (1986) showed that the amount of pores filled by water is directly correlated with available carbon and mineralization potential, thereby indicating the role played by moisture in microbial activity and, consequently, FCO_2. According to Xu and Qi (2001), FCO_2 from pine (*Pinus ponderosa*) plantations is influenced by three main factors: low soil moisture, high temperature, and presence of organic carbon deriving from crop residues.

In addition to physical factors, microbiological factors of soils are of great importance to FCO_2 studies. Microbiological activity related to FCO_2 is influenced by the availability of carbon, nitrogen, phosphorus, and sulfur in the soil and its water content, aeration, pH, and granulometry. According to Jenkinson and Ladd (1981), microbial activity is responsible for organic waste decomposition, nutrient cycling, and energy flow in the soil, thus affecting carbon storage, nutrient availability to plants, and FCO_2. In a study by Xu and Qi (2001), FCO_2 was significantly and positively correlated with microbial biomass, root biomass, and nitrogen, organic matter, and magnesium content.

Thus, when changes in vegetation—caused by anthropogenic activities or improper management practices of agricultural crops—occur, the soil environment is altered, leading to a decline in the stock of organic matter and, consequently, an increase in GHG emission by the carbon content mineralized and transferred from soil to atmosphere in the form of CO_2. Conversely, under appropriate management conditions, the system may sequester CO_2. The high temperatures and humidity levels of Brazil's tropical soils—and, as a result, high rates of organic matter oxidation—constitute favorable conditions for CO_2 production. Because agricultural activity and land use changes account for 22% and 51% of Brazil's total GHG emission, respectively (Intergovernmental Panel Climate Change 2007; Sistema de Estimativas de Emissões de Gases do Efeito Estufa 2018a; 2018b), removal of natural vegetation in Brazilian forests and savannas has caused much concern in the scientific community.

SOIL CO_2 EMISSION IN BRAZIL

An experiment conducted by Dias (2006) in Brazil's Amazon region—in Sinop, MT (11° 24.75' S; 55° 19.50' ° W); Caxiuanã, PA (1° 43' S; 51° 27' W); Manaus, AM (2° 50' S; 60° 0' W), and Santarem, PA (2° 85' S; 54° 95' W)—reported FCO_2 ranging from 0.76 to 12.78 µmol CO_2 $m^{-2}s^{-1}$. Table 1 shows average values for the dry and rainy seasons, respectively: 3.03 µmol CO_2 $m^{-2}s^{-1}$ and 5.76 µmol CO_2 $m^{-2}s^{-1}$ (Sinop); 5.07 µmol CO_2 $m^{-2}s^{-1}$ and 6.09 µmol CO_2 $m^{-2}s^{-1}$ (Caxiuanã);, 5.47 µmol CO_2 $m^{-2}s^{-1}$ and 5.44 µmol CO_2 $m^{-2}s^{-1}$(Manaus), and, 2.90 µmol CO_2 $m^{-2}s^{-1}$ and 5.64 µmol CO_2 $m^{-2}s^{-1}$ (Santarém).

Located in the State of Mato Grosso, Sinop gets 2,000 mm rainfall yearly, with 4 months of dry weather (June through September); its average annual temperature is 24°C, varying slightly according to season. Its vegetation—a transitional tropical forest on Quartzipsamment soil, 423 m above sea level—occupies the ecotone between the Amazon Rainforest and Cerrado. Caxiuanã has an average temperature of 29°C and average annual rainfall of 2,500 mm, with a dry season from June to September and a rainy season from December to March. Its native vegetation is equatorial evergreen rainforest on Oxisols and Gleysols (Dias 2006). In Manaus, the average rainfall is 2,250 mm per year, with a dry season from July to September and a rainy season from December to March. The average annual rainfall in Santarém is 2,190 mm, with a dry season (less than 100 mm of rain per month) from July to November and a 4-month rainy season (December through March). Its average annual temperature is 28°C. The vegetation in both places is equatorial evergreen rainforest on Oxisols (Dias 2006).

Table 1. Comparative averages of $F\text{CO}_2$ from Amazon region

Author	Locality City, State	Average $\mu mol\ CO_2\ m^{-2}s^{-1}$	Methodology
Kleper et al. 1990	Manaus, AM	4.7	IRGA**
Wofsy et al. (1998)	Reserva Ducke, AM	4.5	IRGA*
Meir et al. (1996)	Reserva Jarú, RO	5.5	IRGA*
Trumbore et al. (1995)	Paragominas, PA	6.1	IRGA*
Chamber et al. (2002)	Manaus, AM	3.2	IRGA*
Nunes (2003)	Juruena, MT	4.25	IRGA*
Sotta et al. (2004)	Manaus, AM	6.4	IRGA*
Souza (2004)	Manaus, AM	5.76	IRGA*
Valentine (2004)	Sinop, MT	5.3	IRGA*
Dias (2006)	Sinop, MT	3.03/5.76	IRGA*
Dias (2006)	Caxiuanã, PA	5.07/6.09	IRGA*
Dias (2006)	Manaus, AM	5.47/5.44	IRGA*
Dias (2006)	Santarém, PA	2.90/5.64	IRGA*
Zanchi (2012)	Campina, AM	3.08	IRGA*
Zanchi (2012)	Cuieiras, Am	3.82	IRGA*

Source: Adapted from Nunes (2003), Sotta (2004) and Dias (2006). **static chamber, *dynamic chamber

Significant ($p < 0.05$) correlations were found between $F\text{CO}_2$ and soil moisture in Sinop during its dry season ($R^2 = 0.76$) and rainy season ($R^2 = 0.78$) and in Caxiuanã in the dry season ($R^2 = 0.82$) and rainy season ($R^2 = 0.82$). The same was found in Manaus: significant correlations for the dry ($R^2 = 0.68$) and rainy ($R^2 = 0.60$) seasons (Dias, 2006). Table 1 shows the results found by Dias (2006) and other researchers in Brazil's Amazon region, encompassing six experimental areas in Manaus, with the lowest value (3.2 µmol CO_2 $m^{-2}s^{-1}$) measured by Chamber et al. (2002) and the largest one (6.4 µmol CO_2 $m^{-2}s^{-1}$) by Sotta et al. (2004).

In the Amazon, Dias (2006) found the highest FCO_2 values during the rainy season (November through May) with peaks from February to March. The CO_2 flux in drier areas (Sinop and Santarém) doubled as compared to that in the rainy season. Seasonal variation of soil CO_2 flux in the Amazon region was due to environmental factors, such as temperature and soil moisture, which control FCO_2 in tropical environments and have a direct impact on microbiological activity, organic matter decomposition, and root respiration.

Castellano et al. (2017) quantified FCO_2 in swaths of the Atlantic Forest at different regeneration stages, in two areas located in Rio Claro, São Paulo State, whose climate is classified as Cwa as per the Köppen's system, i.e., mesothermal and tropical altitude. Its average annual temperature is 20.6°C (average above 22°C from December to March, reaching 23°C in February). Its colder months are May through August, with temperatures below 19°C. Its annual rainfall is 1,534 mm, with two well-marked seasons: a rainy season (October through March), with rainfall reaching 1,188 mm (77% of its yearly precipitation rate), and a drier period (April through September), with average rainfall of 346 mm (23% of its yearly precipitation rate). Its FCO_2 ranged from 0.51 to 3.86 µmol CO_2 m^{-2} s^{-1}. The average FCO_2 measured in its regenerated forest and newly-forested areas were 1.91 µmol CO_2 m^{-2} s^{-1} and 1.38 µmol CO_2 m^{-2} s^{-1}, respectively (Table 2). Castellano et al. (2017) indicate that FCO_2 is fairly correlated with soil moisture (r 0.53, p < 0.0001), C/N ratio (r 0.27, p < 0.05), and measurement time (r 0.33, p < 0.05).

Panosso et al. (2009), evaluating FCO_2 at sugarcane (*Saccharum officinarum*) fields under pre-harvesting burning system and green harvesting system, found values of 2.86 and 2.06 µmol CO_2 m^{-2} s^{-1}, respectively. Table 2 also shows a study conducted by Corradi (2003) assessing the correlation between FCO_2 and amount of mulch on the soil: 3.16 µmol CO_2 m^{-2} s^{-1} for soil without mulch, 2.06 µmol CO_2 m^{-2} s^{-1} for soil with 3 t ha^{-1} of mulch, and 2.23 µmol CO_2 m^{-2} s^{-1} for soil with 6 t ha^{-1} of mulch. Tavares (2014) investigated FCO_2 at two sugarcane plantations under green harvesting (5 and 10 years without burning): 2.33 and 2.89 CO_2 μmol CO m^{-2} s^{-1} in the rainy season and 1.19 and 2.62 μmol CO_2 m^{-2} s^{-1} in the dry season. A significant difference between the dry and wet seasons (p > 0.05) was observed in the 5-year green harvesting area: 33% increase in the wet season. Soil temperature is another factor that significantly correlated with FCO_2 in the 5-year green harvesting area (R^2 = 0.80). Notwithstanding, this correlation with soil temperature should be carefully analyzed, since it was influenced by soil moisture at both sites: R^2 = 0.85 (5-year green harvesting area) and R^2 = 0.90 (10-year green harvesting area).

Table 2. Averages of FCO_2 from Atlantic Forest and Cerrado areas

Author	Locality State	Average μmol CO_2 m^{-2}s^{-1}	Methodology
Panosso et al. (2009)	São Paulo	2.06/2.89	IRGA**
Corradi (2003)	São Paulo	3.16	IRGA**
Corradi (2003)	São Paulo	2.06	IRGA**
Corradi (2003)	São Paulo	2.23	IRGA**
La Scala et al. (2000b)	São Paulo	1.46/2.80	IRGA**
Siqueira Neto et al. (2011)	Goiás	0.86	IRGA**
Siqueira Neto et al. (2011)	Goiás	0.69	IRGA**
Siqueira Neto et al. (2011)	Goiás	0.27	IRGA**
Siqueira Neto et al. (2011)	Goiás	0.67	IRGA**
Tavares (2014)	São Paulo	1.19	IRGA**
Tavares (2014)	São Paulo	2.62	IRGA**
Tavares (2014)	São Paulo	2.33	IRGA**
Tavares (2014)	São Paulo	2.89	IRGA**
Castellano (2017)	São Paulo	1.38/1.91	IRGA*
Pereira (2018)	Minas Gerais	2.55	IRGA*
Rigon et al. (2018)	São Paulo	3.6	IRGA**
Rigon et al. (2018)	São Paulo	1.56	IRGA**
Silva et al. (2019)	São Paulo	2.06	IRGA**
Silva et al. (2019)	São Paulo	3.86	IRGA**
Silva et al. (2019)	São Paulo	5.3	IRGA**
Souza et al. (2019)	São Paulo	2.8/3.4	IRGA**

** static chamber, * dynamic chamber

Souza et al. (2019) conducted a study—an experimental design consisting of soils submitted to chiseling in the planting row and in total area—on the correlation between soil carbon dioxide emission and soil pore distribution in Oxisol and Ultisol under sugarcane cultivation. As to Oxisol, the lowest average FCO_2 value was found in soil submitted to chiseling in the planting row (2.8 CO_2 μmol CO m^{-2} s^{-1}) and the highest value (3.4 CO_2 μmol CO_2 m^{-2} s^{-1}) in soil submitted to chiseling in total area, whereas in Ultisol, soil tillage did not affect emissions, indicating that tillage intensity favors FCO_2 in more clayey-textured soils.

In Brazil, GHG emission has been primarily measured in sugarcane plantation areas, due to its importance to biofuel production. Several activities in the sugarcane production process can be considered GHG emission sources, such as land use change, fertilization, organic fertilization, fertigation and irrigation, tillage, sugarcane burning, and diesel use in agricultural operations (Tavares 2014).

Rigon et al. (2018) conducted two experiments in Botucatu, São Paulo State (22°49' S, 48°25' W; 780 m altitude), in Rhodic Hapludox (RH) and Typic Rhodudalf (TR) soils (Soil Survey Staff, 2010), with mesothermal climate and dry austral winters, a well-defined dry season from May to September, and mean annual rainfall of 1,450 mm. FCO_2 was measured during the summer, at soybean (*Glycine max*) plantations after no-till crop rotations with triticale (*Triticum aestivum* x *Secale cereale*), sunflower (*Helianthus annuus*), ruzigrass (*Brachiaria ruziziensis*), grain sorghum (*Sorghum bicolor*), pearl millet (*Pennisetum glaucum*), sunn hemp (*Crotalaria juncea*), and forage sorghum as winter crop rotation systems. The average FCO_2 rate was 3.6 μmol CO_2 m^{-2} s^{-1} and 5.3 μmol CO_2 m^{-2} s^{-1} in TR and RH, respectively. Crop residue content and soil moisture were the main drivers of FCO_2. According to the authors, the use of spring and winter crop rotations may be a viable option for mitigating soil GHG emission.

A study conducted by Pereira (2018) in a native Cerrado area—Sete Lagoas, Minas Gerais State (19° 28'S; 44° 15'W); 732 m altitude; Köppen's climate classification Aw with marked dry season—showed a FCO_2 rate of 2.55 μmol CO_2 m^{-2} s^{-1} in the rainy season and 0.86 μmol CO_2 m^{-2} s^{-1} in the dry season. In addition, an evaluation of monoculture and crop rotation systems in no-till areas indicated lower emission rates for crop rotation as compared to monoculture and native savanna. In general, low emissions occurred during the dry season, mainly influenced by low soil water content. Cumulative FCO_2 decreases after harvest (and during the dry season) by approximately 59% in monocultures, 80% in rotation systems, and 56% in the savanna. In a study at another Cerrado site in the State of Goiás, Siqueira-Neto et al. (2011) found the following average FCO_2 rates: 0.86 µmol CO_2 m^{-2} s^{-1} for native savanna, 0.69 µmol CO_2 m^{-2} s^{-1} for pasture, 0.27 µmol CO_2 m^{-2} s^{-1} for conventional planting with soybean, and 0.67 µmol CO_2 m^{-2} s^{-1} for no-till planting with corn (*Zea mays*).

Silva et al. (2019) investigated temporal variation of FCO_2 in response to three soil tillage operations: intensive soil tillage (rotary hoe+conventional leveling harrow), reduced tillage (disc harrow+leveling harrow), and no tillage. Showing that FCO_2 in the area with intensive soil tillage (3.86 µmol CO_2 m^{-2} s^{-1}) was 87% higher than in the reduced-tillage area (2.06 µmol CO_2 m^{-2} s^{-1}) and 147% higher than in the no-tillage area (1.56 µmol CO_2 m^{-2} s^{-1}). Soil management drives pore class distribution, an important physical attribute to explaining temporal variations of FCO_2.

CONCLUSION

Soil respiration plays an important role in the carbon dioxide cycle of tropical rainforests and savannas, but few studies have been carried out in these biomes. The average FCO_2 rate in Amazon rainforest areas is 4.84 µmol CO_2 m^{-2} s^{-1} (n = 19). Only one study was conducted in the Atlantic forest, reporting an average of 1.9 µmol CO_2 m^{-2} s^{-1} in a reforested area. An experiment in a Cerrado area reports 2.55 µmol CO_2 m^{-2} s^{-1} in the rainy season and 0.86 µmol CO_2 m^{-2} s^{-1} in the dry season, indicating the need for further research on FCO_2 in these biomes. Nonetheless, there are several studies measuring FCO_2 mainly at sugarcane plantations in areas converted for agricultural use.

It was observed that soil moisture and temperature are the factors that most influence FCO_2. However, soil moisture does not always present the same patterns of spatial and temporal variability. In general, soil moisture is negatively correlated with FCO_2 in spatial variability studies and positively in temporal variability ones. The influence of temperature on soil respiration should be carefully analyzed, since soil temperature correlates with soil moisture. It is common to find a positive correlation between moisture and soil temperature in tropical regions, whose dry season is often colder than their wet season and where temperature variation is minor. Consequently, a bivariate model employing temperature and soil moisture may not be capable of explaining temporal variations of FCO_2 under these conditions. Soil respiration may be more efficiently estimated by employing soil moisture in a univariate model (Xu and Qi, 2001; Epron et al., 2004 and 2006; Kosugi et al., 2007; La Scala et al., 2010; Carbonell-Bojollo et al., 2013).

The use of classical statistics in data interpretation presupposes stationarity and independence between samples; however, most soil attributes are spatially dependent (e.g., soil water content and temperature). Assessing this variability provides important information about soil CO_2 dynamics to modelling spatial dependence of gas emissions. Still, the cost of collecting and analyzing samples for mapping specific management sites can be a limiting factor, which explains why spatial variability is not studied in forest and savanna biomes (Kang et al., 2003; Epron et al., 2004; Epron et al., 2006; Teixeira et al., 2011; Brito et al., 2009; Panosso et al., 2009; Montanari et al., 2005).

Therefore, spatial variability analysis studies on agricultural crops have been designed to get a better understanding of these aspects. In addition, multivariate analysis is an important tool for joint data analysis, especially in order to identify possible correlations between soil attributes. The spatial heterogeneity of FCO_2 is related to root biomass and microbial biomass, amount of crop residue, organic carbon, and nitrogen, cation exchange capacity, density, porosity, acidity, and topographic position (Epron et al., 2004; Brito et al., 2009; Teixeira et al., 2013). Temporal variability can be attributed to variations in soil temperature or humidity (Panosso et al., 2008; Carvalho et al., 2010; Maier et al., 2010; Goutal et al., 2012). However, the premise that soil respiration can be represented by linear relationships—

even when several variables are taken into account—is not supported by the literature.

Several research projects have been designed to measure soil respiration in Brazil, but most of them evaluate agricultural areas rather than native biomes. While studies in natural areas focus on temporal variability, in agricultural areas, the emphasis is on spatial variability, which can be explained by higher operational costs and greater difficulties in setting up plots to study geoestatistcs in forests and savannas due to the presence of bushes and trees and long distances for collecting samples. Considering the associated uncertainties—due to the large number of variables that may influence emission, different sampling methodologies, and small number of studies in each biome—, it is not possible to compare the absolute values of the research projects developed in Brazil aimed at characterizing FCO_2.

REFERENCES

Beuchle, R., Grecchi, R. C., Shimabukuru, Y. E., Seliger, R., Eva, H. D., Sano, E. & Achard, F. (2015). "Land cover changes in the Brazilian Cerrado and Caatinga biomes from 1990 to 2010 based on a systematic remote sensing sampling approach." *Applied Geography*, 58, 116-127. doi: 10.1016/j.apgeog.2015.01.017.

Bustamante, M. M. C., Martinelli, L. A., Ometto, J. P. H. B., Carmo, J. B., Jaramillos, V., Gavitos, M., Araujo, P., Austin, A. T. A., Perez, T. & Marquina, S. (2014). "Innovations for a sustainable future: rising to the challenge of nitrogen greenhouse gas management in Latin America." *Current Opinion in Environmental Sustainability*, 9, 73-81. doi: 10.1016/j.cosust.2014.09.002.

Campanili, Maura. & Wigold, B. Schaffer. (2010). *Mata Atlântica: patrimônio nacional dos brasileiros*. Brasília: Ministério do Meio Ambiente. [*Atlantic Forest: National Heritage of all Brazilians*. Brasília: Ministry of the Environment].

Carbonell-Bojollo, R., Torres, M. A. R. R., Rodriguez-Lizana, A. & Ordonez-fernandez, R. (2012). "Influence of soil and climate conditions on CO_2 emissions from agricultural soils." *Water Air Soil Pollution, 223*(6), 3425-3435. Doi: 10.1007/s11270-012-1121-9.

Cardoso, E. L., Silva, M. L. N., Silva, C. A., Curi, N. & Freitas, D. A. F. (2010). "Estoques de carbono e nitrogênio em solo sob florestas nativas e pastagens no bioma do pantanal." *Pesquisa agropecuária brasileira, 45*(9), 1028-1035. ["Carbon and nitrogen stocks in soil under native forests and pastures in the Pantanal biome." *Brazilian Agricultural Research, 45*(9), 1028-1035]. doi: 10.1590/S0100-204X2010000900013.

Castellano, G. R., Moreno, L. X., Menegário, A. A., Govone, J. S. & Gastmans, D. (2017). "Quantificação das emissões de CO_2 pelo solo em áreas sob diferentes estádios de restauração no domínio da Mata Atlântica." *Quimica Nova, 40* (4), 407-412. ["Quantification of soil CO_2 emissions in two forested areas at different regeneration stages in the Atlantic Forest." *New Chemistry, 40* (4), 407-412]. doi:10.21577/0100-4042.20170036.

Corradi, M. M., Panosso, A. R., Martins-Filho, M. V. & La Scala, Jr. N. (2013). "Crop residues on short-term CO_2 emissions in sugarcane production areas." *Engenharia Agrícola, 33*(4), 699-708. doi: 10.1590/S0100-69162013000400009.

Davidson, Eric A., Susan, E. Trumbore. & Ronald, G. Amundson. (2000). "Biogeochemistry - Soil warming and organic carbon content." *Nature, 408* (6814), 789-790. doi: 10.1038/35048672.

Davidson, Eric A. & Wayne, T. Swank. (1986). "Environmental parameters regulating gaseous nitrogen losses from two forested ecosystems via nitrification and denitrification." *Applied and Environmental Microbiology, 52*(6), 1287-1292. Accessed January 14, 2020. https://www.ncbi.nlm.nih.gov/pmc/articles/PMC239223/.

Denman, K. L., Brasseur, G., Chidthaisong, A., Ciais, P., Cox, P. M., Dickinson, R. E., Hauglustaine, D., Heinze, C., Holland, E., Jacob, D., Lohmann, U., Ramachandran, S., da Silva Dias, P. L., Wofsy, S. C. & Zhang, X. (2007). "Couplings Between Changes in the Climate System and Biogeochemistry." In *Climate Change 2007: The Physical Science Basis. Contribution of Working Group I to the Fourth Assessment*

Report of the Intergovernmental Panel on Climate Change, edited by S. Solomon, D. Qin, M. Manning, Z. Chen, M. Marquis, K. B. Averyt, M. Tignor and H. L. Miller, 500-587. Cambridge, UK and New York, USA: Cambridge University Press. Accessed January 14, 2020. https://www.ipcc.ch/site/assets/uploads/2018/ 02/ar4-wg1-chapter7-1.pdf.

Dias, Jadson Dezincourt. (2006). *Fluxo de CO_2 proveniente da respiração do solo em áreas de floresta nativa da Amazônia*. Dissertação de mestrado, Universidade de São Paulo. [*CO_2 flux from soil respiration in native Amazon forest areas.* Master's thesis, University of São Paulo]. Accessed January 14, 2020. https://www.teses.usp.br/teses/ disponiveis/ 91/91131/tde-04102006-163445/publico/JadsonDias.pdf.

Dixon, R. K., Brown, S., Houghton, R. A., Solomon, A. M., Trexler, M. C. & Wisniewski, J. (1994). "Carbon pools and flux of global forest ecosystems." *Science*, *263*(5144), 185-190. doi:10.1126/science. 263.5144.185.

Epron, D., Bosc, A., Bonal, D. & Freycon, V. (2006). "Spatial variation of soil respiration across a topographic gradient in a tropical rain forest in French Guiana." *Journal of Tropical Ecology*, *22*(5), 565-574. doi: 10.1017/S0266467406003415.

Epron, D., Nouvellon, Y., Roupsard, O., Mouvondy, W., Mabiala, A., Saintandre, L., Jofre, R., Jourdan, J., Bonnefond, J. M., Berbigier, P. & Hamel, O. (2004). "Spatial and temporal variations of soil respiration in a eucalyptus plantation in Congo." *Forest, Ecology and Management*, *202*(1-3), 149-160. doi:10.1016/j.foreco.2004.07.019.

Fang, C., Moncrieff, J. B., Gholz, H. L. & Clark, K. L. (1998). "Soil CO_2 efflux and its spatial variation in a Florida slash pine plantation." *Plant and Soil*, *205*(2), 135-146. Accessed January 14, 2020. http://www.paper.edu.cn/scholar/showpdf/MUj2QN5IOTD0gxeQh.

Feigl, Brigitte J., Jerry M. Melillo & Carlos C. Cerri. (1995). "Change the origin and quality of soil organic matter after pasture introduction in Rondônia (Brazil)." *Plant and Soil*, *175*, 227-241. doi: 10.1007/BF02413007.

Fernandes, S. A. P., Bernoux, M., Cerri, C. C., Feigl, B. J. & Piccolo, M. C. (2002). "Seasonal variation of soil chemical properties and CO_2 and CH_4 fluxes in unfertilized and P- fertilized pastures in an Ultisol of the Brazilian Amazon." *Geoderma*, *107* (3-4), 227-241. doi: 10.1016/ S0016-7061(01)00150-1.

Food and Agriculture Organization of the United Nations. (2001). *State of the World's Forests: 2001*. Rome: Food and Agriculture Organization. Accessed January 14, 2020. http://www.fao.org/3/y0900e/ y0900e 00.htm.

Food and Agriculture Organization of the United Nations. (2016). *2016 The state of food and agriculture: Climate change, agriculture and food security*. Rome: Food and Agriculture Organization. Accessed January 14, 2020. http://www.fao.org/3/a-i6030e.pdf.

Forster, Helmut W. & Antonio Carlos G. de Mello. (2007). "Biomassa aérea de raízes em arvores de reflorestamento heterogêneo no vale do Paranapanema, SP." *Instituto Florestal - Série Registro*, *31*, 153-157. ["Aerial root biomass in heterogeneous reforestation trees in the Paranapanema Valley, SP." *Forest Institute - Record Series*, *31*, 153-157]. Accessed January 14, 2020. https:// smastr16.blob.core. windows.net/iflorestal/RIF/SerieRegistros/IFSR31/IFSR31_153-157.pdf.

Goreau, Thomas J. & William, Z. de Mello. (1987). "Effect of deforestation on sources and sinks of atmospheric carbon dioxide, nitrous oxide, and methane from some central Amazonian soils and biota." In *Proceedings of the Workshop on biogeochemistry of tropical rain forest: problems for research*, edited by D. Athie, T. E. Lovejoy, and P. M. Oyens. Piracicaba: FEALQ.

Goutal, N., Parent, F., Bonnaud, P., Demaison, J., Nourisson, G., Epron, D. & Ranger, J. (2012). "Soil CO_2 concentration and efflux as affected by heavy traffic in forest in northeast France." *European Journal of Soil Science*, *63*, 261-271. doi:10.1111/J.1365-2389.2011.01423.X.

Houghton, R. A., Skole, D. L., Nobre, C. A., Hackler, J. L., Lawrence, K. T. & Chomentowski, W. H. (2000). "Annual fluxes of carbon from deforestation and regrowth in the Brazilian Amazon." *Nature*, *403*, 301-304. doi: 10.1038/35002062.

Intergovernmental Panel Climate Change. (2001). "Climate Change 2001: Impacts, Adaptation and Vulnerability." In *Contribution of Working Group II to the Third Assessment Report of the Intergovernmental Panel on Climate Change*, edited by J. J. McCarthy, O. F. Canziani, N. A. Leary, D. J. Dokken, and K. S. White. Cambridge, UK, and New York, USA: Cambridge University Press. Accessed January 14, 2020. https://library.harvard.edu/collections/ipcc/docs/27_WGIITAR_FINAL.pdf.

Jenkinson, David S. & Jeffrey N. Ladd. (1981). "Microbial biomass in soil: measurement and turnover." In *Soil biology and Biochemistry*, *5*, edited by E. A. Paul, and J. N. Ladd, 415-471. New York: Marcel Dekker.

Kang, S., Doh, S., Lee, D. S., Lee, D., Jin, V. L. & Kimball, J. S. (2003). "Topographic and climatic controls on soil respiration in six temperate mixed-hardwood forest slopes, Korea." *Global Change Biology*, *9*, 1427-1437. doi:10.1046/j.1365-2486.2003.00668.x.

Keller, Michael., Warren, A. Kaplan. & Steven, C. Wofsy. (1986). "Emission of N_2O, CH_4 and CO_2 from tropical forest soils." *Journal of Geophysical Research Atmospheres*, *91*(11), 1791-1802. doi: 10.1029/JD091iD11p11791.

Kosugi, Y., Mitani, T., Itoh, M., Nogushi, S., Tani, M., Matsuo, N., Takanashi, S., Ohkubo, S. & Nik, A. R. (2007). "Spatial and temporal variation in soil respiration in a Southeast Asian tropical rainforest." *Agricultural and Forest Meteorology*, *147*, 35-47. doi: 10.1016/j.agrformet.2007.06.005.

Kuntoro, Arno A. & Ade Wahyu. (2009). "The Effect of Deforestation on Regional Terrestrial Carbon Balance: A Case Study of Borneo Island." *Journal of International Development and Cooperation*, *15*(1-2), 141-165. Accessed January 14, 2020. https://pdfs.semanticscholar.org/beb3/c2992f32168328d5db2738ffdb71b6fe8b63.pdf.

Kutsch, Werner L., Michael Banh & Andreas Heinemeyer. (2010). *Soil Carbon Dynamics: an integrated methodology*. Cambridge: Cambridge University Press.

Lapola, D. M., Martinelli, L. A., Peres, C. A., Ometto, J. P. H. B., Ferreira, M. E., Nobre, C. A., Aguiar, A. P. D., Bustamante, M. M. C., Cardoso, M. F., Costa, M. H., Joly, C. A., Leite, C. C., Moutinho, P., Sampaio, Strassburg, B. B. N. & Vieira, I. C. G. (2013). "Pervasive transition of the Brazilian land-use system." *Nature Climate Change*, *4*, 27-35. doi: 10.1038/nclimate2056.

La Scala, Jr. N., Marques, J., Pereira, G. T. & Cora, J. E. (2000a). "Short-term temporal changes in the spatial variability model of CO_2 emissions from a Brazilian bare soil." *Soil Biology & Biochemistry*, *32*(10), 1459-1462. doi: 10.1016/S0038-0717(00)00051-1.

La Scala, Jr. N., Marques, J., Pereira, G. T. & Cora, J. E. (2000b). "Carbon dioxide emission related to chemical properties of a tropical bare soil." *Soil Biology and Biochemistry*, *32*, 1469-1473. doi: 10.1016/S0038-0717(00)00053-5.

La Scala, Jr. N., Mendonça, E. S., Souza, J. J., Panosso, A. R., Simas, F. N. B. & Schaefer, C. E. G. R. (2010). "Spatial and temporal variability in soil CO_2-C emissions and relation to soil temperature at King George Island, Maritime Antarctica." *National Institute of Polar research*, *4*(3), 479-487. doi: 10.1016/j.polar.2010.07.001.

Le Dantec, Valerie, Daniel Epron & Erick Dufrene. (1999). "Soil CO_2 efflux in beech forest: comparison of two closed dynamic systems." *Plant and Soil*, *214*(1), 125- 132. doi: 10.1023/A:1004737909168.

Malhi, Yadvinder, Dennis D. Baldocchi & Paul G. Jarvis. (1999). "The carbon balance of tropical, temperate and boreal forest." *Plant, Cell and Environment*, *22*, 715-740. Accessed January 14, 2020. https://nature.berkeley.edu/biometlab/pdf/mahli_pce_1998%2022_715.pdf.

Medina, E., Klinge, H., Jordan, C. F. & Herrera, R. (1980). "Soil respiration in Amazonian rain forests in the Rio Negro Basin." *Flora*, *170*(3), 240–250. doi: 10.1016/S0367-2530(17)31209-4.

Meir, P., Grace, J., Miranda, A. & Lloyd, J. (1996). "Soil respiration in Amazônia and in cerrado in central Brazil." In *Amazonian deforestation and climate*, edited by J. C. H. Gash, C. A. Nobre, J. M. Roberts, and R. Victoria, 578-595. Chichester, New York, Brisbane, Singapore: John

Wiley & Sons. Accessed January 14, 2020. http://marte3.sid.inpe.br/col/sid.inpe.br/iris@1905/2005/07.29.21.29.20/doc/6046.pdf.

Ministério da Ciência, Tecnologia e Inovação. (2014). "Estimativas anuais de emissões de gases de efeito estufa no Brasil." 2nd Ed. Brasilia: Minístério da Ciência, Tecnologia e Inovação. ["Annual estimates of greenhouse gas emissions in Brazil." 2nd Ed. Brasilia: Ministry of Science, Technology and Innovation]. Accessed January 14, 2020. http://sirene.mctic.gov.br/portal/export/sites/sirene/backend/galeria/arquivos/2018/10/11/Estimativas_2ed.pdf.

Montanari, R., Marques, Jr. J., Pereira, G. T. & Souza, Z. M. (2005). "Forma da paisagem como critério para otimização amostral de latossolos sob cultivo de cana-de-açúcar." *Pesquisa Agropecuária Brasileira*, 40, 69-77. ["Landscape form as a criterion for sampling optimization of an oxisol under cultivation of sugarcane." *Brazilian Agricultural Research*, 40, 69-77]. Accessed January 14, 2020. http://www.scielo.br/pdf/pab/v40n1/23244.pdf.

Nunes, Paulo César. (2003). *Inluência do efluxo de CO_2 do solo na produção de forragem numa pastagem extensiva e num sistema agrosilvopastoril.* Dissertação de mestrado, Universidade Federal de Mato Grosso. [*Influence of soil CO_2 efflux on forage production in extensive pasture and agro-forestry-pastoral system.* Master's dissertation, Federal University of Mato Grosso, Cuiabá].

Pacto pela restauração da Mata Atlântica. *"O Pacto." Pacto pela restauração da Mata Atlântica.* Abril 7, 2009. *["The Pact." Pact for the restoration of the Atlantic Rainforest.* April 7, 2009]. Assessed December 25, 2019. https://www.pactomataatlantica.org.br/o-pacto.

Panosso, A. R., Marques, Jr. J., Pereira, G. T. & La Scala, Jr. N. (2009). "Spatial and temporal variability of soil CO_2 emission in a sugarcane area under green and slash-and-burn managements." *Soil and Tillage Research*, 105, 275-282. doi: 10.1016/j.still.2009.09.008.

Panosso, A. R., Pereira, G. T., Marques, Jr. J. & La Scala, Jr. N. (2008). "Variabilidade espacial da emissão de CO_2 em Latossolos sob cultivo de cana-de-açúcar em diferentes sistemas de manejo." *Engenharia Agrícola*, 28(2), 227-236. ["Spatial variability of CO_2 emission in

Oxisol soils cultivated with sugarcane under different management practices." *Agricultural Engineering*, *28*(2), 227-236]. doi: 10.1590/S0100-69162008000200003.

Redo, Daniel, Mitchell Aide, T. & Matthew L. Clark. (2013). "Vegetation change in Brazil's dryland ecoregions and the relationship to crop production and environmental factors: Cerrado, Caatinga, and Mato Grosso, 2001–2009." *Journal of Land Use Science*, *8*(3), 123- 153. doi:10.1080/1747423X.2012.667448.

Rigon, G., Calonego, J. C., Rosolem, C. A. & La Scala, Jr. N. (2018). "Cover crop rotations in no-till system: short-term CO_2 emissions and soybean yield." *Scientia agricola*, *75* (1), 1-16. doi:10.1590/1678-992x-2016-0286.

Rodrigues, Ricardo R. (1999). "A vegetação de Piracicaba e os municípios do entorno." *Circular técnica IPEF*, *89*, 1-17. ["The vegetation of Piracicaba and surrounding municipalities." *Technical publishing IPEF*, *89*, 1-17]. Accessed January 14, 2020. https://www.ipef.br/publicacoes/ctecnica/nr189.pdf.

Sabine, C. L., Feely, R. A., Gruber, N., Key, R. M., Lee, K., Buloister, J. L., Wanninkhof, R., Wong C. S., Wallace D. W. R., Tilbrook, B., Millero, F. J., Peng, T., Kozyr, A. & Rios, A. F. (2004). "The oceanic sink for anthropogenic CO_2." *Science*, *305*, 367-371. doi:10.1126/science.1097403.

Schlesinger, William H. (1997) *Biogeochemistry; analysis of global change.* 2nd ed. Oxon: Academic Press.

Siqueira Neto, M., Piccolo, M. D. C., Costa, Jr. C., Cerri, C. C. & Bernoux, M. (2011). "Emissão de gases do efeito estufa em diferentes usos da terra no bioma Cerrado." *Revista Brasileira de Ciência do Solo*, *35*(1), 63-76. ["Greenhouse gas emission caused by different land-uses in Brazilian savanna." *Brazilian Journal of Soil Science*, *35*(1), 63-76]. doi: 10.1590/S0100-06832011000100006.

Silva, B. O., Moitinho, M. R., Santos, G. A. A., Teixeira, D. B., Fernandes, C. & La Scala, Jr. N. (2019). "Soil CO_2 emission and short-term soil pore class distribution after tillage operations." *Soil and Tillage Research*, *186*, 224-232. doi: 10.1016/j.still.2018.10.019.

Sistema de Estimativas de Emissões de Gases do Efeito Estufa. (2018a). *Emissões do Setor de Agropecuária.* Brasil: Observatório do Clima, IMAFLORA. [*Agricultural Sector Emissions.* Brazil: Climate Observatory, IMAFLORA.] Accessed January 14, 2020. https://www.imaflora.org/downloads/biblioteca/Relatorios_SEEG_2018-Agro_Final_v1.pdf.

Sistema de Estimativas de Emissões de Gases do Efeito Estufa. (2018b). *Emissões do Setor de Mudança de Uso da Terra.* Brasil: Observatório do Clima, IMAZON, IPAM. [*Emissions from the Land Use Change sector.* Brazil: Climate Observatory, IMAZON, IPAM]. Accessed January 14, 2020. http://www.observatoriodoclima.eco.br/wp-content/uploads/2018/05/Relato%CC%81rios-SEEG-2018-MUT-Final-v1.pdf.

Souza, L. M., Fernandes, C., Moitinho, M. R., Bicalho, E. S. & La Scala, Jr. N. (2019). "Soil carbon dioxide emission associated with soil porosity after sugarcane field reform." *Mitigation and Adaptation Strategies for Global Change,* 24(1), 113–127 doi: 10.1007/s11027-018-9800-5.

Soil Survey Staff. (2010). *Keys to Soil Taxonomy.* 11th Ed. Washington: USDA-Natural Resources Conservation Service.

Sotta, Eleneide Doff. (1998). *Fluxo de CO_2 entre solo e atmosfera em floresta tropical úmida da Amazônia Central.* Dissertação de Mestrado, Instituto Nacional de Pesquisas da Amazônia, 1998. [*CO_2 flux between soil and atmosphere in the tropical rainforest of Central Amazonia.* Master's Dissertation, National Amazon Research Institute]. Accessed January 14, 2020. https://ainfo.cnptia.embrapa.br/ digital/bitstream/item/28955/1/Sotta-1998-DissMestrado.pdf.

Tavares, Rose Luiza Moraes. (2014). *Emissão de CO_2 e atributos físicos, químicos e biológicos do solo em sistemas de manejo de cana-de-açúcar.* Tese de Doutorado, Universidade Estadual de Campinas. [*CO_2 Emission and Physical, Chemical and Biological Attributes of Soil in Sugarcane Management Systems.* PhD diss., Campinas State University]. Accessed January 14, 2020. http://repositorio.unicamp.br/bitstream/REPOSIP/257129/1/Tavares_RoseLuizaMoraes_D.pdf.

Teixeira, D. B., Panosso, A. R., Cerri, C. E. P., Pereira, G. T. & La Scala, Jr. N. (2011). "Soil CO_2 emissions estimated by different interpolation techniques." *Plant and Soil, 345,* 187-194. doi: 10.1007/s11104-011-0770-6.

Word Metereological Organization. (2019). "The State of Greenhouse Gases in the Atmosphere Based on Global Observations through 2018." *WMO Greehouse Gas Bulletin, 15,* 1-8. Accessed January 14, 2020. https://library.wmo.int/doc_num.php?explnum_id=10100.

Xu, Ming. & Ye, Qi. (2001). "Soil surface CO_2 efflux and its spatial and temporal variations in a young ponderosa pine plantation in northern California." *Global Change Biology, 7,* 667-677. doi: 10.1046/j.1354-1013.2001.00435.x.

Zanchi F. B., Waterloo, M. J., Kruijt, B., Kesselmeier, J., Luizão· F. J., Manzi, A. O. & Dolman, A. J. (2012). "Soil CO_2 efflux in central Amazonia: environmental and methodological effects." *Acta Amazonica, 42* (2), 0. doi: 10.1590/S0044-59672012000200001.

In: Carbon Dioxide Emissions
Editor: Asia Santana

ISBN: 978-1-53617-763-3
© 2020 Nova Science Publishers, Inc.

Chapter 3

BRAZIL AT COP21: CHALLENGES TO ACHIEVE CARBON EMISSION REDUCTION TARGETS

Marcelo Silva Sthel, Marcenilda Amorim Lima and Fernanda Gomes Linhares*

Center of Science and Technology,
North Fluminense State University,
Campos dos Goytacazes, Brazil

ABSTRACT

In December 2015, the United Nations Framework Convention on Climate Change (UNFCCC) held the Conference of the Parties (COP21). Brazil has committed to reduce Greenhouse Gasses (GHGs) emissions in 37% by 2025, and 43% by 2030, based on 2005 levels. The current Brazilian policy for deforestation control in the Amazon region is in line with the emission-reducing promises made at COP21. Brazil has programs for reducing GHGs via Intended Nationally Determined Contributions (INDC), by transforming the land use and the forestry sector. The country has made a pledge to adopt additional measures in this sector so as to

---

* Corresponding Author's E-mail: sthel@uenf.br.

achieve the goals agreed in Paris, such as: strengthening compliance with the Brazilian Forestry Code at national, state and city levels; supporting policies to achieve zero illegal deforestation in the Brazilian Amazon by 2030; restoring and reforesting 12 million hectares of forests by 2030, for multiple uses; and scaling up sustainable native forest management systems through georeferencing and traceability applicable to native forest management, in order to discourage illegal and unsustainable practices. In January 2019, a new government took position in Brazil by promising to change the Brazilian environmental policy. In the following August, the National Space Research Institute (INPE) released data on Amazon deforestation, which indicated a considerable growth when compared with 2018, thus contradicting the additional measures proposed by the Brazilian government. This fact created an international crisis between Brazil and the G7 summit European countries. In this way, in case the Amazon deforestation increase continues, will it be possible to fulfill the Brazilian Paris agreement goals?

Keywords: Paris Agreement, carbon emission, climate change

INTRODUCTION

The Paris climate conference COP21 comprised the significant participation of 195 countries, the greatest attendance progress in 20 years of climate conferences (UNFCCC, 2015; Tollefson and Weiss, 2015). The objective was to obtain commitments from all participating parties, which would guarantee a cutback in greenhouse gas emissions (GHG). This would limit the global warming rate to below 2°C, equivalent to pre-industrial levels, and would ensure efforts to keep it at 1.5°C (Fawcett et al., 2015; Hulme, 2016). Until November 2016, the agreement had been ratified by 100 countries, corresponding to 69% of the world's GHG output (UNFCCC, 2016). In November 2017, Syria was the last country to sign the Paris agreement (Harvey, 2017) despite the occurring civil war at that time. Hence, the USA became the only country out of the accord, and shall remain so based on President Donald Trump's directives. A scathing criticism of the Paris Accord comes from the climate scientist James Hansen (The New York Times, 2015; Hansen et al., 2017), who considers the emission reducing goals proposed by the countries to be inadequate, for China is allowed to continue to increase its emissions until 2030. The contrast in the Earth's energy balance led Hansen et al., (Hansen et al., 2008; 2013; 2016) to recommend a reduction of CO_2 emissions to the level of 350 ppm, which implies a global temperature close to current values, that is about 1°C in relation to pre-industrial era. The monthly CO_2 concentration in the Mauna Loa was 403.64 ppm in October 2017 (NOAA, 2017), which represented a concentration greater than that recommended by James Hansen in order to avoid the great environmental impacts on the planet.

Jackson et al., (2017) indicated there would be a 2% increase in emissions in 2017, compared to the 2016. In 2018, the possibility of a continuous increase on account of the growth projections in the world economy is witnessed. This poses enormous challenges in keeping the global average temperature increase below 2°C. Peters et al., (2017) and Le Quéré et al., (2017) also indicate an increase in global emissions of 2% when considering a 3.5% increase in China's emissions, an increase of 2% in India's, a decrease of 0.4% in the USA's and a 1.9% increase in emissions regarding the rest of the world. The economic dynamics is different among the ratifying countries of the Paris agreement, as they generate differing emission rates. The current trend of increases in emission levels is a matter of concern and could seriously compromise the COP21 targets set for 2025 and 2030.

Brazil ratified the Paris treaty in September 2016 (TOLLEFSON, 2016), committing to reducing its GHG emissions in 37% by 2025 and in 43% by 2030, based on 2005 levels, as for instance the Intended Nationally Determined Contributions, INDCs (Brazil, 2016). According to data from the Greenhouse Gas Emissions Estimate System (SEEG, 2019) from the Climate Observatory, Brazil emitted 2.2 billion tons of CO_2eq in 2016. This represented an increase of 8.9% compared to 2015, being the highest emission rate since 2008. Brazil is now the world`s seventh largest emitter of CO_2, for its emissions increased by 12.3% between 2015 and 2017 despite the economic recession, which produced a fall of 7.4% in the Gross Domestic Product (GDP). This is mainly due to agricultural activities and changes in land use, which represents 74% of Brazil's total emissions. Figure 1 shows the emissions in giga tons of CO_2eq regarding Brazil's various sectors between 2004 and 2017 (SEEG, 2019). The energy sector, on the other hand, presented a decrease in its emissions. For example, electricity generation was responsible for 30% less emissions in 2016 given both the economic recession and the recurrence of rainfall. Contrastingly, during the period of 2011 to 2015, the water crisis led to higher emissions, mainly because of the increase in the use of thermoelectric plants (Mendes and Sthel, 2017).

The 2026 decennial Energy Plan for Brazil prepared by the Energy Research Company (EPE, 2017) predicts, for the next decade, that 70.5% of all investment in energy will be in fossil fuels. The Brazilian Federal Chamber has recently voted the Provisional Measure 795 that expands the fossil fuels (Militão, 2017) subsidy, which may produce a fiscal waiver of up to 300 billion dollars in 25 years. The government's proposal of tax exemption for offshore oil (Watts, 2017) increases future emissions, contradicting the country's progressive position during the Bonn climate conference. For attitudes such as these, Brazil has earned the "Fossils of the Day" award at the COP23 (350, 2017). Meanwhile, the Norwegian Sovereign Fund for oil and gas (Vaughan, 2017) has evaluated the possibility of reducing investments in fossil fuels on account of the price variations related to these finite sources in recent years. Christophe McGlade and Paul Ekins (McGlade and Ekins, 2015) report that greenhouse gas emissions present in current fossil fuel global reserves estimates are around three times higher. Hence, the unabated use of all current fossil fuel reserves is incompatible with a warming limit of 2°C. The fiscal incentive policies for fossil fuels increase the risks of failing to meet the safe limits for rising temperatures this century.

The United Nations Environment Program (UNEP, 2017) published "The Emissions Gap Report 2017" about the humanity's climate debt. It points out that the period from 2018 to 2020 is the last chance for the stabilization of global warming by 2°C, or ideally by 1.5°C, thus increasing the targets of the INDCs proposed nowadays. However, the so-called Talanoa Dialogue (Facilitating Dialogue), which took place at a meeting with COP21 participating countries in 2018, discussed the possibility of broadening the targets agreed for reducing emissions and increasing funding for environmental programs. As Patrick T. Brown and Ken Caldeira (Brown and Caldeira, 2017) put it, "our results suggest that achieving any given global temperature stabilization target will require steeper greenhouse gas emissions reductions than previously calculated."

Therefore, it is important to discuss, in 2019, the expansion of the emission reduction targets. With the current policy of supporting fossil sources, without urgent measures to reduce deforestation, and with unsustainable practices in its agriculture and livestock sector, Brazil will find it difficult to scale up its emissions reduction targets as established at the Talanoa Dialogue. Thus, the country will need to revise its emission reduction targets between 2018 and 2020, as well as carry out a critical assessment on the possibility of achieving those targets already ratified in 2016 (Tollefson, 2016).

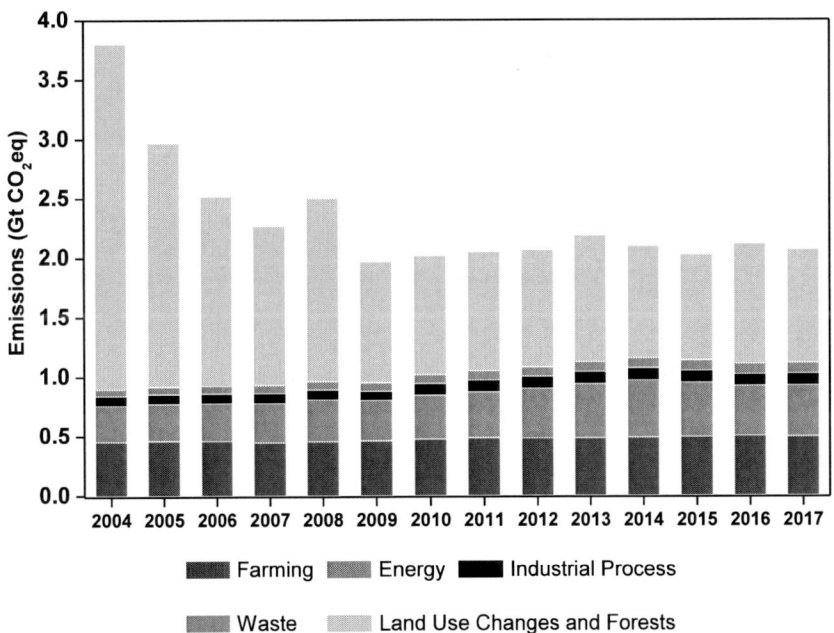

Figure 1. Emissions of CO_2eq considering the various sectors in Brazil per year (SEEG, 2019).

Forest burning (Sexton et al., 2016; Wingfield et al., 2015; Rowland et al., 2015; Trumbore, Brando and Hartmann, 2015; Bello, et al., 2015; Lewis et al., 2015;) impacts human health, biodiversity and climate, which is contrary to all global climate policy, and has turned difficult to meet the Paris agreement's targets globally. According to Bastin et al.:

> Global trees restoration is the most effective solution to climate change to date. The global canopy cover scenario may shrink 223 million hectares by 2050, most of the losses will occur in tropical forests. The results highlight the opportunity for climate change mitigation through global tree restoration, but it also requires urgent action". As Senior et al.,[35] stated, "as Range shifts are a crucial mechanism enabling species to avoid extinction under climate change. The majority of terrestrial biodiversity is concentrated in the tropics, including species considered most vulnerable to climate warming, but extensive and ongoing deforestation of tropical forests is likely to impede range shifts. We conduct a global assessment of the potential for tropical species to reach analogous future climates - 'climate connectivity'- and empirically test how this has changed in response to deforestation between 2000 and 2012. We find that over 62% of tropical forest area (~10 million km^2) is already incapable of facilitating range shifts to analogous future climates. 12 years, continued deforestation has caused a loss of climate connectivity. Limiting further forest loss and focusing the global restoration agenda towards creating climate corridors are global priorities for improving resilience of tropical forest biotas under climate change (Bastin et al., 2019).

Various authors (Kelley et al., 2019; Andela et al., 2017 and FORKET et al., 2017) show new mathematical models and satellite systems to understand the dynamics of burned forests, and how humans alter global fire regimes. Although vegetation fires affect human infrastructure, ecosystems, global vegetation distribution and climate composition, the climatic, environmental and socioeconomic factors that control global fire activity in vegetation are poorly understood, and, notwithstanding, they have various complexities in their formulations, which are represented in global fire - vegetation model processes. Data-based approaches, such as machine learning algorithms, were proposed for better identifying and understanding the control factors for fire activity. Satellite observations can predict fire activity, as the SOFIA version 1 approach for instance. In SOFIA, the models use various variables and their functional relations to estimate the burned area, and it can be easily adapted to the burning models on most complex vegetation's. Furthermore, they result in higher performance regarding the burned area-related prediction. The importance of the Amazon is unquestionable for human society, influencing on the global climate for the immense biodiversity it possesses. Hence, its conservation is vital for Brazil in order to meet the goals agreed at COP21.

DEFORESTATION IN THE BRAZILIAN AMAZON

The deforestation in the Amazon grew 27% in 2016, compared to 2015, and as announced by the Brazilian government, deforestation dropped a 16% in 2017, compared to 2016. In 2018, it had an increase of 8.5% (7,536 km²) compared to 2017 (6,947 km²). It also represents a 73% reduction compared to 2004 (INPE, 2019), when the Federal Government launched the Amazon Deforestation Prevention and Control Plan (PPCDAm). It is noted there had been a considerable reduction in deforestation in the Amazon region between 2004 and 2014. The Figure 2 shows the deforested Amazon areas in km², based on data from the National Institute for Space Research (INPE), depicting also an increase in deforestation from 2015 on.

Deforestation decrease data are in line with data on CO_2eq emission reductions from the same period (Figure 1), showing a direct relationship between the Amazon deforestation rates and the carbon emissions rates. Table 1 (INPE, 2019) shows a comparison of the deforestation of the Brazilian Amazonian states in km² between the years of 2017 and 2018; in the latter year, the states of Pará and Mato Grosso were the most deforested with 2744 km² and 1490 km² respectively. Pará had a deforestation increase of 13% compared to 2017, whereas Mato Grosso did it by 5%. Figures 3a and 3b show MCO_2eq emissions in Pará and Mato Grosso states between 2014 and 2017. Figures 3c and 3d show a reduction in deforestation in the states of Pará and Mato Grosso from 2004 to 2018, thus indicating a direct correlation between the reduction in CO_2eq emissions and the decrease in deforestation rates in these states.

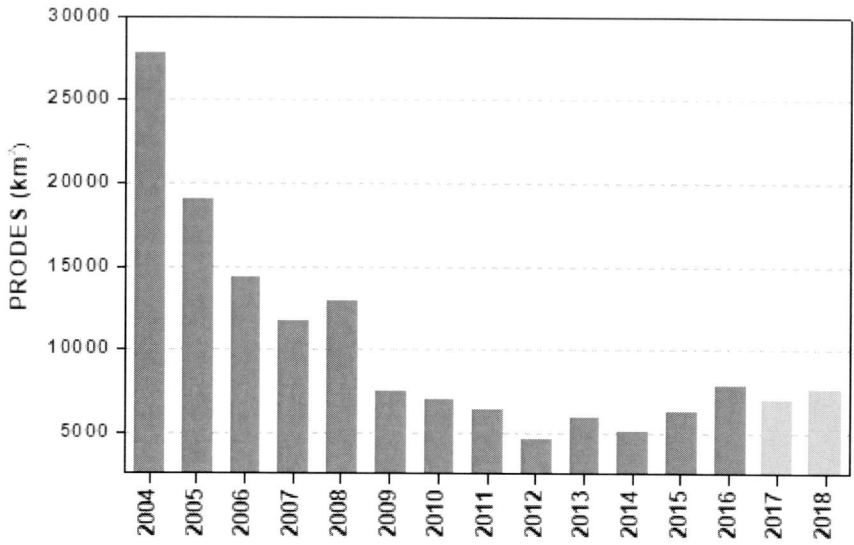

Source: INPE, 2019.

Figure 2. Annual deforestation rate in the Legal Amazon (km²).

Table 1. Comparison of deforestation rates of 2017 and 2018

State	PRODES 2017 (km²)	PRODES 2018 (km²)	Contribution (%)
Acre	257	444	73
Amazonas	1,001	1,045	4
Amapá	24	24	0
Maranhão	265	253	-5
Mato Grosso	1,561	1,490	-5
Pará	2,433	2,744	13
Rondônia	1,243	1,316	6
Roraima	132	195	48
Tocantins	31	25	-19
Total	6,947	7,536	8

Source: INPE, 2019.

Rate PRODES Amazônia - 2004 and 2018 (km^2)

In January 2019, the new Brazilian government who took office brought along a political discourse not favorable to environmental issues, which pointed out new guidelines for the Amazon region, namely the use of indigenous lands for mining and expansion of areas for agribusiness, for example. In August 2019, it was reported that Brazil had had more than 72,000 fire outbreaks to that time in the year, an increase of 84% over the same period in 2018 (INPE, 2019). Fire outbreaks have also been confirmed by the American Space Agency (NASA, 2019). A notable increase in large, intense and persistent fires is perceived from burnings on the main roads of central Brazilian Amazon, as explained by Douglas Morton, head of NASA's Goddard Space Flight Center Biospheric Science Laboratory: "Satellites are often the first to detect fires in remote regions of the Amazon." NASA's main fire detection tool, since 2002, has been the MODIS (Moderate Resolution Image Spectroradiometer) instrument on the Terra and Aqua satellites.

Morton (in NASA, 2019) noted that fire activity statistics from 2019 distributed by NASA and Brazil's National Institute for Space Research (INPE) are congruent. Actually, "INPE also uses active fire data from NASA's MODIS sensors to monitor fire activity in the Brazilian Amazon," said Morton. "As a result, NASA and INPE have the same estimates of changes in recent fire activity. MODIS detections are higher in 2019 than at the time of last year in the seven states that make up the Brazilian Amazon. At this point in the fire season, active fire detections by MODIS in 2019 are higher in the Brazilian Amazon than in any year since 2010." Upon confirming such data, the German Prime Minister Angela Merkel backed the request from the French President Emmanuel Macron to insert the Brazilian deforestation on the G7 summit agenda, held in late August, indicating thus an international environmental crisis.

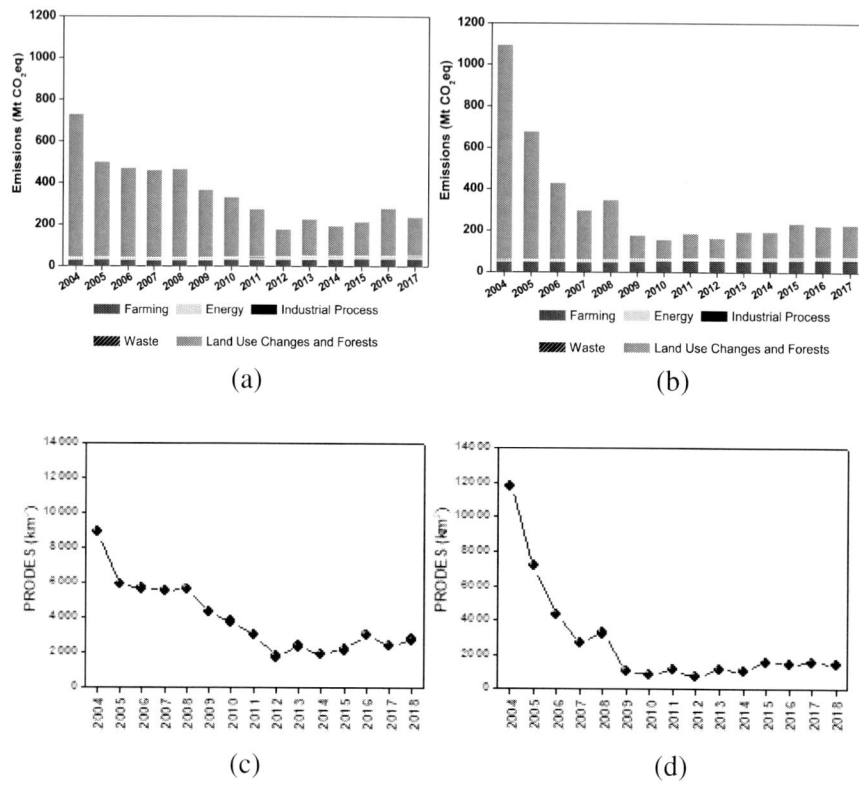

Figure 3. Greenhouse gas emissions from different sectors, between 2004-2017: a) State of Pará; b) State of Mato Grosso. Source: SEEG, 2019. Amazon deforestation rate PRODES from 2004 to 2018 (km^2): c) State of Pará; d) State of Mato Grosso. Source: INPE[39].

Table 2. Alerts (Deter) x deforestation rate (Prodes)

Years	Deter (km^2)	Prodes (km^2)
2018/2019	6,833.9	-
2017/2018	4,572	7,536
2017/2016	4,639.2	6,947
2016/2015	5,378	7,893

Source: INPE (Oliveira, 2019).

Brazilian President Jair Bolsonaro responded vehemently, "I regret that President Macron seeks to take advantage of what is a national issue in Brazil and other Amazonian countries for personal political gains," calling it a "sensationalist tone." UN Secretary-General Antonio Guterres said he was quite concerned, for "in the midst of the global climate crisis, we cannot afford more damage to a major source of oxygen and biodiversity." This crisis gained prominence in the international press (Phillips, 2019; BBC, 2019; Samuel, 2019; Nossiter, 2019; Andrade, 2019; Gauchazh, 2019) with a worldwide repercussion, especially in Brazil, where it was centered on an intense political debate. Bolsonaro, in the face of international controversy, has authorized the deployment of the armed forces to fight fires in the Amazon (MAZVI et al., 2019). On August 19th, São Paulo, the largest city in South America, had its skies darkened in the middle of the afternoon (The Economist, 2019; Prizibisczki, 2019). According to INPE researcher Alberto Setzer (Prizibisczki, 2019), "this event is a combination of two coincident but distinct physical conditions: the entry of a cold air front, and the presence of a cloud of smoke from burnings that originated hundreds and even thousands of miles away." They are particles from fires that occurred in the Midwest and North regions, between Paraguay and Mato Grosso, which encompass parts of Bolivia, Mato Grosso do Sul and Rondônia. Some of these fires occurred almost 3000 km away (Barnes, 2019).

The Real-Time Deforestation Detection System (DETER) is a rapid (daily) deforestation alert system that increases surveillance efficiency by monitoring via satellite where changes in vegetation occur, such as partial or total tree cuts. Official deforestation data are provided by the Brazilian Satellite Amazon Forest Monitoring Program (PRODES), which issues an annual deforestation assessment, with a 95% reliability index (Diniz et al., 2015). INPE has released data for the annual Deter system series from August 2018 to July 2019 (Oliveira, 2019) (Figure 4), indicating a total deforestation of 6,833 km^2. This figure is 50% higher when compared to the period from August 2017 to July 2018, when it reached 4,572 km^2. Table 2 shows a comparison between the Deter deforestation alert forecast and the Prodes deforestation official data forecast for the 2015/2016, 2016/2017, 2017/2018 and 2018/2019 biennia, the latter and current is yet to be fully accounted (Oliveira, 2019). However, if we compare this current biennium with the 3 previous biennia, there is a possibility for Prodes to indicate a deforested area of more than 10,000 km^2 for 2019.

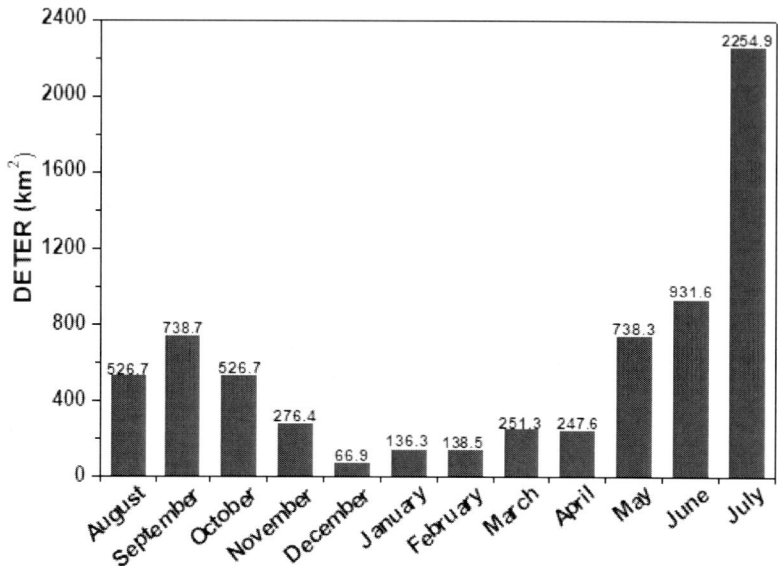

Source: Deter/TerraBrasilis/Inpe (Oliviera, 2019).

Figure 4. Deter alerts, monthly, on the vegetation variation-capturing system, which releases suspected acts of deforestation in the Legal Amazon, from August 2018 to July 2019.

From 2018 to 2019, the largest number of warnings came from the state of Pará, with the following lands under alert: 98.8 km² in the regions of Jamanxim National Forest; 55.8 km² in the Tapajós Environmental Protection Area; and 24.7 km² in the Terra do Meio Ecological Station. The dataset confirms that this state is the most threatened within the Amazon region, mainly on account of the extensive livestock and export agriculture.

CONCLUSION

Amid the international polemic caused by the Amazon deforestation growth in 2019, compared to 2018, the statements from the Brazilian president combined with the subsequent halt in the money transfers to the Amazon fund by Norway (Boffey, 2019) further aggravate the problem, for this fund's main objective is to fight forest destruction and support sustainable economic activities. The wildfires in the Amazon are one of the main causes of increased carbon emissions countrywide. Brazil has programs for reducing GHGs by means of the Intended Nationally Determined Contributions (INDC), by transforming the land use and the forestry sector, such as strengthening compliance with the Forest Code at the federal, state and municipal levels; strengthen policies and measures to achieve zero illegal deforestation in the Brazilian Amazon by 2030, and offset greenhouse gas emissions derived from suppressing legal vegetation; restore and reforest 12 million hectares of forests for multiple uses; expand the scale of sustainable native forest management tools via georeferencing and traceability systems applicable to this field, with a view to discouraging illegal and unsustainable practices. In the agricultural sector, we find the attempt to strengthen the Low Carbon Agriculture Plan (ABC Plan) (Brasil, 2012) as the main strategy for sustainable development in agriculture, including the additional restoration of 15 million hectares of degraded pasture, and the increase of 5 million hectares of crop-livestock-forest integration (iLPF) systems by 2030. In the face of the current events, it becomes challenging for Brazil to meet the targets agreed at COP21. The measures proposed by the current government are not effective for forest protection, which lead to an international concern about the real possibility of an increase in the Amazon forest burning rates.

REFERENCES

Andela, N.; Morton, D. C.; Giglio, L.; Chen, Y.; Van Der Werf, G. R.; Kasibhatla, P. S.; Defries, R. S.; Collatz, G. J.; Hantson, S.; Kloster, S.; Bachelet, D.; Forrest, M.; Lasslop, G.; Li, F.; Mangeon, S.; Melton, J. R.; Yue, C.; Randerson, J. T. (2017). A human-driven decline in global burned area. *Science* 356, 1356-362.

Andrade, Rodrigo de O. (2019). Alarming surge in Amazon fires prompts global outcry. *Nature.* doi: 10.1038/d41586-019-02537-0 (Published in: August 23, 2019). https://www.nature.com/articles/ d41586-019-02537-0.

BBC. (2019). News: Amazon fires: Markel and Macron urge G7 to debate 'emergency'. (Published in August 23, 2019). https://www.bbc.com/news/world-latin-america-49443389.

Barnes, Angela. (2019). Amazon burning: Smoke travels nearly 3000 km to black out São Paulo in middle of the day. *EuroNews.* Published in August 23, 2019. https://www.euronews.com/2019/08/21/amazon-burning-sao-paulo-s-smoke-filled-skies-make-it-dark-at-2-p-m.

Bastin, Jean-Francois; Finegold, Yelena; Garcia, Claude; Mollicone, Danilo; Rezende, Marcelo; Routh, Devin; Zohner, Constantin M.; Bello, Carolina; Galetti, Mauro; Pizo, Marco A.; Magnago, Luiz Fernando S.; Rocha, Mariana F.; Lima, Renato A. F.; Peres, Carlos A.; Ovaskainen, Otso; Jordano, Pedro. (2015). Defaunation affects carbon storage in tropical forests. *Science Advance* 1,1-10.

Boffey, Daniel. (2019). Norway halts Amazon fund donation in dispute with Brazil. *The Guardian. Brussels.* Published in August 18, 2019. https://www.theguardian.com/world/2019/aug/16/norway-halts-amazon-fund-donation-dispute-brazil-deforestation-jair-bolsonaro

Brasil (2012). Plano setorial de mitigação e de adaptação às mudanças climáticas para a consolidação de uma economia de baixa emissão de carbono na agricultura: Plano ABC (Agricultura de Baixa Emissão de Carbono) / Ministério da Agricultura, Pecuária e Abastecimento, Ministério do Desenvolvimento Agrário, coordenação da Casa Civil da Presidência da República. – Brasília: MAPA/ACS, 173 p. ISBN 978-85-7991-0620. [*Sectoral climate change mitigation and adaptation Plan for the consolidation of a Low Carbon economy in agriculture*: ABC Plan (Low Carbon Agriculture) / Ministry of Agriculture, Livestock and Supply, Ministry of Agrarian Development, coordination of the Civil house of the Presidency of the Republic. – Brasília: MAPA/ACS, 173 p. ISBN 978-85-7991-0620].

Brazil, Federative Republic (2016). *Intended Nationally Determined Contribution Towards Achieving the Objective of the United Nations Framework Convention on Climate Change.* http://www.itamaraty.gov.br/images/ed_desenvsust/BRAZIL-iNDC-english.pdf * (accessed in 6 November 2017).

Brazilian Space (2019). *INPE consolidates 7536 km² of Amazon deforestation in 2018* (Published in June 25, 2019). https://brazilianspace.blogspot.com/2019/06/inpe-consolida-7536-km-de-desmatamento.html

Brown, Patrick T. and Caldeira, Ken. (2017). Greater future global warming inferred from Earth's recent energy budget. *Nature* 552, 45-50.

Crowther, Thomas W. (2019). The global tree restoration potential. *Science* 365, 76-79.

Diniz, Cesar G.; Souza, Arleson A. A.; Santos, Diogo C.; Dias, Mirian C.; Luz, Nelton C.; Moraes, Douglas R. V.; Maia, Janaina S.; Gomes, Alessandra R.; Narvaes, Igor Da S.; Valeriano, Dalton M.; Maurano, Luis E. P. And Adami, Marcos. Deter-B: The New Amazon Near Real-Time Deforestation Detection System. *IEEE Journal of selected topics in applied Earth observations and remote sensing*, V. 8, N. 7, 3619-3628 (2015).

EPE. (2017). *Empresa de Pesquisa Energética: Plano Decenal de Expansão de Energia 2026*. http://www.epe.gov.br/pde/Paginas/default.aspx (accessed in December 3, 2017) [*Energy Research Company: Ten-Year Energy Expansion Plan 2026*].

Fawcett, Allen A. et al. (2015). Can Paris pledges avert severe climate change? Reducing risks of severe outcomes and improving chances of limiting warming to 2°C. *Science* 350, 1168-1169.

Forkel, Matthias; Dorigo, Wouter; Lasslop, Gitta; Teubner, Irene; Chuvieco, Emilio and Thonicke, Kirsten. (2017). A data-driven approach to identify controls on global fire activity from satellite and climate observations (SOFIA V1). *Geosci. Model Dev* 10, 4443-4476.

Gauchazh. (2019). *Capa de revista "The Economist" cita Amazônia sob o risco de morte e pede vigília global*. Published in August 01, 2019. https://gauchazh.clicrbs.com.br/ambiente/noticia/2019/08/capa-da-revista-the-economist-cita-amazonia-sob-risco-de-morte-e-pede-vigilia-global-cjysvcku300kf01ph7q2ybocn.html. [*Magazine cover The Economist quote Amazon at risk of death and calls for global vigil*. Published in August 01, 2019].

Hansen, James et al. (2017). Young people's burden: requirement of negative CO2 emissions. Earth Syst. Dynam., 8, 577-616.

Hansen, James et al. (2008). Target atmospheric CO2: Where should humanity aim?, Open Atmos. Sci. J., 2, 217–231.

Hansen, James et al. (2013). Assessing "dangerous climate change": Required reduction of carbon emissions to protect young people, future generations and nature, Plos One, 8(12) e81648.

Hansen J. et al. (2016). Evidence from paleoclimate data, climate modeling, and modern observations that 2 ₒC global warming could be dangerous, Atmos. Chem. Phys., 16, 3761–3812, 2016, https://doi.org/10.5194/acp-16-3761,2016.

Harvey, Fiona (2017). *Syria signs Paris climate agreement and leaves US isolated.* https://www.theguardian.com/environment/2017/nov/07/ syria -signs-paris-climate-agreement-and-leaves-us-isolated (Accessed in November 29, 2017).

Hulme, Mike (2016). 1.5 °C and climate research after the Paris Agreement. *Nature Climate Change* 6, 222-224.

INPE. (2019). *Coordenação Geral de Observação da Terra: INPE consolida 7536 km² de desmatamento na Amazônia em 2018.* http://www.obt.inpe.br/OBT/noticias/inpe-consolida-7-536-km2-de-desmatamento-na-amazonia-em-2018. (Published in: Junho 25, 2019). [*General Earth Observation Coordination: INPE consolidates 7536 km² of Amazon deforestation in 2018*].

INPE (2019). *PRODES - Amazônia: Monitoramento do Desmatamento da Floresta Amazônia Brasileira por Satélite.* http://www.obt.inpe.br/ OBT/assuntos/programas/amazonia/prodes. [*Amazon: Brazilian Amazon Rainforest Deforestation Monitoring by Satellite*].

Jackson, R. B.; Le Quéré, C; Andrew, R. M.; Canadell, J. G.; Peters, G. P.;Roy, J. and Wu, L. (2017). Warning signs for stabilizing global CO2 emissions. *Environ. Res. Lett.* 12, 110202.

Kelley, Douglas I.; Bistinas, Ioannis; Whithey, Rhys; Burton, Chantelle; Marthews, Toby R. and Dong, Ning. (2019). How contemporary bioclimatic and human controls change global fire regimes. *Nature Climate Change* 9, 690-696.

Le Quéré, C. et al. Global Carbon Budget 2017. *Earth Syst. Sci. Data Disc* https://doi.org/10.5194/essd-2017-123 (2017)

Lewis, Simon L.; Edwards, David P.; Galbraith, David. (2015). Increasing human dominance of tropical forests. *Science* 349, 827-832.

Mazui, Guilherme; Barbiéri, Luiz Felipe and Garcia, Gustavo. Bolsonaro autoriza uso das forças armadas no combate a queimadas na Amazônia. G1: *Política. Brasília.* Published in August 23, 2019. https://g1.globo.com/politica/noticia/2019/08/23/bolsonaro-autoriza-uso-das-forcas-armadas-no-combate-a-queimadas-na-amazonia.ghtm. [Bolsonaro authorizes the use of the armed forces to combat burning in the Amazon. G1: *Politics. Brasília.* Published in August 23, 2019].

Mcglade, Christophe and Ekins, Paul. (2015). The geographical distribution of fossil fuels unused when limiting global warming to 2 oC. *Nature* 517, 187-190.

Mencuccini, M. and Meir. P. (2015). Death from drought in tropical forests is triggered by hydraulics not carbon starvation. *Nature* 528, 119-122.

Mendes, Luiz Fernando Rosa and Sthel, Marcelo Silva. (2017). *Thermoelectric Power Plant for Compensation of Hydrological Cycle Change: Environmental Impacts in Brazil. Case Studies in the Environment,* 1-7, https://doi.org/10.1525/cse.2017.000471.

Militão, Eduardo. (2017). Estudos apontam perda de R$ 1 trilhão em renúncia fiscal após leilão do pré-sal. *UOL: Economia.* Published in October 31, 2017. https://economia.uol.com.br/noticias/redacao/2017/10/31/estudos-apontam-perdas-de-r-1-tri-em-renuncia-fiscal-com-leilao-do-pre-sal.htm (accessed in December 3, 2017). [Studies show loss of $ 1 trillion in tax waiver after pre-sal auction. *UOL: Economy.* Published in October 31, 2017].

NASA. Earth Observatory. (2019). Uptick in Amazon fire activity in 2019. Story by Adam Voiland, with information from Douglas Morton. https://earthobservatory.nasa.gov/images/145498/uptick-in-amazon-fire-activity-in-2019.

NOAA (Earth System Research Laboratory) (2017). *Trends in Atmospheric Carbon Dioxide.* https://www.esrl.noaa.gov/gmd/ccgg/trends/global.html. (last access: December 01, 2017).

Nossiter, Adam. (2019). *The New York Times: How Emmanuel Macron positioned himself as star of the G7 show.* (August 27, 2019) ttps://www.nytimes.com/2019/08/27/world/europe/g7-macron-trump-france.html.

Oliveira, Elida. (2019). *Balanços oficiais de desmatamento da Amazônia confirmam dados de sistema de alerta; entenda.* G1. Published in August 18, 2019. https://g1.globo.com/natureza/noticia/2019/08/18/balancos-oficiais-de-desmatamento-da-amazonia-confirmam-dados-de-sistema-de-alerta-entenda.ghtml. [amazon official deforestation balance sheets confirm alert system data; understand. G1. Published in August 18, 2019].

Peters, G. P. et al. Towards real-time verification of CO_2 emissions. (2017). *Nature Climate Change* 7, 848-852.

Phillips, Tom. (2019). *The Guardian: Merkel backs Macron's call for G7 talks on Amazon fires.* (Published August 23, 2019). https://www.theguardian.com/world/2019/aug/23/amazon-rainforest-fires-macron-calls-for-international-crisis-to-lead-g7-discussions.

Prizibisczki, Cristiane. (2019). *Pesquisadores descrevem fenômeno que escureceu SP e sua relação com as queimadas na Amazônia.* O Eco. Published in August 22, 2019. https://www.oeco.org.br/noticias/pesquisadores-descrevem-fenomeno-que-escureceu-sp-e-sua-relacao-com-as-queimadas-na-amazonia/. [Researchers describe a phenomenon that darkened SP and its relationship with the burning in the Amazon. O Eco. Published in August 22, 2019].

Rowland, L.; Costa, A. C. L.; Galbraith, D. R.; Oliveira, R. S.; Binks, O. J.; Oliveira, A. A. R.; Pullen, A. M.; Doughty, C. E.; Metcalfe, D. B.; Vasconcelos, S. S.; Ferreira, L. V.; Malhi, Y.; Gracel, J.; Trumbore, S.; Brando, P.; Hartmann, H. (2015). Forest health and global change. *Science* 349, 814-818.

Samuel, Henry. (2019). Amazon fires: Brazilian president sends army to tackle blaze after Emmanuel Macron moves to block EU-South America trade deal. *The Telegraph.* (Published in August 24, 2019) https://www.telegraph.co.uk/news/2019/08/22/brazilian-president-says-country-lacks-money-fight-amazon-fires/.

SEEG. (2019). *Sistema de Estimativa de Emissão de Gases.* http://plataforma.seeg.eco.br/total_emission (accessed in October 3, 2019). [*Gas Emission Estimation System*].

Senior, A.; Hill, Jane K. and Edwards, David P. (2019). Global loss of climate connectivity in tropical forests. *Nature Climate Change* 9, 623-626.

Sexton, O.; Noojipady, Praveen; Song, Xiao-Peng; Feng, Min; Song, Dan-Xia; Kim, Do-Hyung; Anand, Anupam; Huang, Chengquan,; Channan, Saurabh; Pimm, Stuart L. And Townshend, John R.. (2016) Conservation policy and the measurement of forests, *Nature Climate Change* 6, 192-196

The Economist. (2019). Forest fires in the Amazon blacken the sun in São Paulo. Published in August 22, 2019. https://www.economist.com/the-americas/2019/08/22/forest-fires-in-the-amazon-blacken-the-sun-in-sao-paulo.

The New York Times (2017). The Road to a Paris Climate Deal. https://www.nytimes.com/interactive/projects/cp/climate/2015-paris-climate-talks/fabius-bangs-his-gavel-and-leaders-cheer (Accessed in November 29, 2017)

Tollefson, Jeff. (2016). Brazil ratification pushes Paris climate deal one step closer. *Nature.* (Published in September 14, 2016) http://www.nature.com/news/brazil-ratification-pushes-paris-climate-deal-one-step-closer-1.20588 (accessed in November 6, 2017)

Tollefson, Jeef and Weiss, Kennethr (2015). Nations adopt historic global climate accord. *Nature* 528, 315-316

UNEP - United Nations Environment Programme (2017). *The Emissions Gap Report 2017.* https://wedocs.unep.org/bitstream/handle/20.500.11822/22070/EGR_2017.pdf (accessed in December 3, 2018)

UNFCCC - Unites Nations - Framework Convention Climate Change (2016). *Paris Agreement - Status of Ratification* (Published in 4 November 2016) http://unfccc.int/paris_agreement/items/9444.php (accessed in November 6, 2016).

UNFCCC - United Nations - Framework Convention on Climate Change (2015). *Conference of the Parties* (COP21). Adoption on the Paris Agreement, Paris.

Vaughan, Adam (2017). World's biggest sovereign wealth fund proposes ditching oil and gas holdings. *The Guardian.* https://www.theguardian.com/business/2017/nov/16/oil-and-gas-shares-dip-as-norways-central-bank-advises-oslo-to-divest (accessed in December 3, 2017).

Watts, Jonathan. (2017). *The Guardian: Brazil's oil sale plans prompt fears of global fossil fuel extraction race.* (Published in November 15, 2017) https://www.theguardian.com/environment/2017/nov/15/brazils-oil-sale-plans-undermine-its-role-at-bonn (accessed in December 3, 2017).

Wingfield, M. J.; Brockerhoff, E. G.; Wingfield, B. D.; Slippers, B. (2015). Planted forest health: The need for a global strategy. *Science* 349, 832 - 836.

350 (2017). Brazil wins 'Fossil of the Day' at COP23. (Published in November 16, 2017) https://350.org/brazil-wins-fossil-of-the-day-at-cop-23/ (accessed in December 3, 2017).

In: Carbon Dioxide Emissions
Editor: Asia Santana

ISBN: 978-1-53617-763-3
© 2020 Nova Science Publishers, Inc.

Chapter 4

ESTIMATES OF THE INFLATION EFFECT OF A GLOBAL CARBON PRICE ON CONSUMER, INVESTMENT, EXPORT AND IMPORT PRICES

Fredrik N. G. Andersson[*]
Department of Economics, Lund University, Lund, Sweden

ABSTRACT

This chapter considers the potential inflation effects of a global carbon price on consumer prices, investment prices, export prices, and import prices. We estimate the effects under three different scenarios. The results clearly indicate that the inflation effects in developed countries of a 100 USD/ton carbon price are small. For developing countries, the inflation effect is larger—potentially too large for it to be politically feasible to introduce a global carbon price. However, a simple adjustment of the price based on the price level in each country equalizes the inflation effects across all countries. In light of this kind of adjustment, a global carbon price is more likely to be implemented.

Keywords: carbon price, inflation, consumer prices, export prices, import prices, investment prices, climate change

---

[*] Corresponding Author's E-mail: fredrik_n_g.andersson@nek.lu.se.

INTRODUCTION

Combating global climate change is a key policy challenge. From a theoretical economic point of view, a carbon price is one of the most efficient and important policy tools available to reduce emissions. The price should be applied uniformly across all economic sectors and all countries for it to be as effective as possible (Weitzman, 2014). Either a national or a regional carbon price, rather than a global price, may lead to outsourcing of carbon-intensive production to locations with either no or a low carbon price and thus have little effect on global emissions (Cole, 2004; Andersson, 2018).

Although carbon pricing has become increasingly popular across regions (Lo, 2012; Newell et al., 2013) and countries (see for example a review by Fedor, 2016), there is still no agreement among countries on a global price. Developing countries, in particular, are concerned that they will be disproportionally negatively affected by such an agreement, because their economies tend to be relatively less carbon-efficient compared to more developed economies. China, for example, emits up to seven times more carbon dioxide (CO_2) per unit GDP compared to the European Union (EU) (World Development Indicators, 2017).

The purpose of this chapter is to study how a global carbon price may affect consumer prices, investment prices, government consumption prices, export prices, and import prices. Previous studies have focused on effects on prices in individual markets, such as prices for cement or steel (e.g., Smale et al., 2006). The effect of a carbon price on aggregate price indices are also important, not least from a policy perspective. An increase in the overall price level will reduce economic welfare. Aggregate prices thus provide an indication of whether a global carbon price is politically feasible. The larger the estimated price effect, the greater the political resistance against the tax will be.

A global carbon price is likely to increase prices. By how much depends on how consumers and firms respond to the carbon price. According to economic theory, a carbon price causes consumers and firms to reduce their consumption of fossil fuels by shifting to alternative fuels and reducing energy waste. A lower consumption of fossil fuels reduces the carbon price effect on aggregate prices (Aldy et al., 2012; Pearce, 1991). Consequently, to estimate the inflation effect of a carbon price we must model how consumers and firms will change their behavior due to the carbon price. In addition, the introduction of a carbon price is often coupled with additional tax reforms such as a cut in value-added tax or employer fees to reduce any potential inflation effect of the carbon price. Consequently, the predicted inflation effect of a carbon price varies from study to study depending on the assumptions made. Bosquet (2000), for example, found that the estimated consumer price effect from 66 different models ranged from a small fall in consumer prices to a moderate increase. A more recent review by Laing et al., (2014) focusing specifically on electricity prices and the price of industrial goods similarly found that the price effect varied from model to model due to the various assumptions made.

An alternative approach to trying to estimate the exact inflation effects using a model is to use scenarios. Unlike models, the purpose of scenarios is to not to predict the exact effect of a policy on the economy. The purpose of scenarios is to map different possible outcomes by altering the assumptions on how the economy will change following a change in policy (Farber et al., 2007). Hence, scenarios provide an overview of the range of possible outcomes.

In this study, we use three specific scenarios to map the the potential effect of a global carbon price on aggregate prices. Our analysis allows us to discuss (i) to what degree technological change alone is sufficient to limit the inflation effects of the carbon price, (ii) to what extent changes in consumption patterns are needed as a complement to the technological changes (iii) how different countries are affected, and (iv) whether a global carbon price is politically feasible to implement. Included in our study are the United States (US), the European Union (EU), other developed countries, and two rapidly growing emerging economies, namely China and India.

The rest of the chapter is organized as follows. In Section 2, we present our method and data. In Section 3, we present the empirical results. Section 4 concludes the chapter.

METHOD

Different Model Approaches

A global carbon price constitutes a major change in policy. How firms and consumers respond to the policy determines what kind of effect it will have on emissions and economic welfare. It is likely that the price will trigger changes in the economy that reduce the use of fossil fuels, but the exact change in behavior and technology is uncertain. There are, broadly speaking, two approaches to model the effects: one is to use models to predict the behavior of consumers and firms and, from those predictions, estimate the effects on, e.g., consumer prices, investment prices, etc. The other approach is to use scenarios with different assumptions as to how consumers and firms will behave to map various potential outcomes based on assumed behavior.

The model approach to predict behavior is based either on a pure economic model or an integrated economic–energy model that also includes the energy system. Broadly speaking, there are three types of integrated models: bottom-up models, top-down models, and models that combine the two approaches (for a review, see Hourcade and Jaccard, 2006; Böhringer and Rutherford, 2008; Hardt and O'Neill, 2017). Bottom-up models focus primarily on the energy system and potential energy carriers and contain fewer details on the economy. Top-down models, on the other hand, have a greater focus on the economy and on how agents in the economy react to various climate policies and contain fewer details on the energy system. Bottom-up models tend to use over-simplified assumptions in relation to the economy, and top-down models tend to use over-simplified assumptions in relation to technology and the energy system. The third category of models, hybrid models, combines both bottom-up and top-down approaches to make the models more realistic.

The response of consumers and firms is modeled using a general equilibrium model. To yield predictions of the future, the models are parametrized using historical outcomes (i.e., estimates of past behavior). The accuracy of the predictions depends both on how realistic the model is and to what degree past behavior is a useful guide for future behavior.

An often-overlooked factor that also affects the accuracy of the models' predictions is "radical uncertainty" about the future (King, 2016). There will be future events that are impossible to forecast in advance and that will have profound effects on the economic outcome. Some of these unforeseeable events will be random, while others will be potentially triggered by a major change in policy. Large policy changes commonly lead to unpredictable changes in behavior, which makes past behavior a poor guide for the future (e.g., Lucas, 1976; Lubik and Surico, 2010).[1]

One approach to deal with radical uncertainty is to use scenarios. The purpose of scenarios is not to try to predict the actual outcome of the economy, as such predictions are difficult to make given the uncertainty concerning how firms and consumers will behave. The purpose of scenarios is to explore contrasting outcomes using different sets of assumptions (Peterson et al., 2003). In other words, the aim of scenarios is not to predict the actual outcome or to paint an exact picture of the future. Instead, scenarios aim to map as many realistic outcomes as possible (Faber et al., 2007, van Vuuren et al., 2011).

Scenarios are, thus, widely used to model possible future economic outcomes under radical uncertainty. In climate economics, scenarios are used to deal with, for example, uncertainty over how the climate will respond to various levels of CO_2 concentrations, how a rise in temperature will affect economic activity, and which technologies which will become available in the future (e.g., Moss et al., 2010, van Vuuren et al., 2011). It is less common to use scenarios to alter the effect of how consumers and firms will change their behavior (e.g., Carpos et al., 2014).

Given the high level of uncertainty regarding households' and firms' responses to a carbon price, here we rely on scenarios to map possible responses of agents to the global carbon price, rather than using a model that

---

[1] Great effort has gone into improving the model to overcome this problem, including improving the model's microeconomic foundations. However, this approach has not solved the instability problem arising from changes in behavior (Oliner et al., 1996; Callabero, 2010; Hurtado, 2014).

attempts to predict such responses based on past behavior. Consequently, we do not employ a model to predict behavior. Instead, we map possible outcomes by assuming alternative possible changes in behavior due to the carbon price. Our study is thus a complement, not a supplement, to studies based on integrated economic–energy models.

Assumptions and Scenarios

We consider three different scenarios. In each scenario, we vary one assumption, which allows us to map the possible responses of the economy to the carbon price.[2] In one of the scenarios we alter the assumption on technology, and in another scenario we alter the implementation of the price. We do not consider a specific scenario including changes in consumption patterns. However, we do discuss how consumption patterns may change and their potential effects on the inflation effects of the carbon price based on our results.

We consider the inflation effect on consumer prices, government consumption prices, investment prices, export prices, and import prices. For consumer prices, we also consider the inflation effect on various groups of consumption goods and services. This decomposition allows us to study the potential distributional effects of a carbon price among consumers within a country, as low-income households generally have higher shares of spending on items such as food and clothing and lower shares of spending on services. These distributional effects are important, though often forgotten in the analysis (Dennig et al., 2015). Here, we can also study the potential effects of a lower pass-through rate in certain sectors.

The three scenarios are:

- *Scenario 1*: a global carbon price of 100 USD/ton CO_2 is introduced worldwide. Countries that already have a put a price on carbon, either through a carbon price or a trading system, are assumed to

---

[2] Whether the price is imposed through a tax or an emissions' trading system has no effect on our results. What matters is the price level, not how the price is imposed.

increase the present price level by 100 USD/ton CO_2. This assumption implies that countries that are already leading other countries in terms of pricing carbon continue to have a more ambitious climate policy.
- *Scenario 2*: Developing countries are often less productive compared to developed countries. This is one of the reasons why the price level is lower in developing countries than in developed countries (Balassa, 1964; Samuelson, 1964; Ravallion, 2013). Thus, 100 USD has greater purchasing power in China and India compared to either the US or the EU. Moreover, a common global carbon price would be a greater economic burden in developing countries compared to developed countries due to the lower price level. Thus, in the second scenario, we adjust the carbon price to the price level. In developed countries, the price is 100 USD. According to data from the the OECD, in 2008, the Chinese price level was 45% of the US price level, and the Indian price level was 28% of the US price level. Thus, the carbon price is set to 45 USD/ton CO_2 in China and 28 USD/ton CO_2 in India in this scenario.
- *Scenario 3*: One effect of a carbon price is that it increases incentives to adopt the latest existing and most carbon-efficient technology. Developing countries, in particular, have a lot to gain by shifting to more carbon-efficient means of production. In Scenario 3, we assume that all countries become as carbon-efficient as the US. In other words, we assume that countries shift to already existing technologies. We do not assume that there is any improvement compared to the levels that already exist today. For some countries, the shift to US levels of carbon efficiency implies a worsening and not an improvement of the carbon efficiency of the economy. This is true for some smaller developed countries. However, for simplicity, we use the US as the benchmark in these calculations.

In all scenarios, the estimated price effect depends partly on the pass-through rate, i.e., how much of the increase in cost is pushed on to the consumers in terms of higher prices. De Buyn et al., (2015) found that the pass-through rates onto the consumers after the EU Emissions Trading Scheme was introduced were close to 100% in the most carbon-intensive sectors.[3] The pass-through rate was particularly high in carbon-intensive industries such as the utilities and metals industries, which are characterized by relatively large actors and limited competition. Similarly, Fabra and Reguant (2014) found that emission costs were almost fully passed on to consumers in the electricity sector in the US, a result confirmed for Australia by Nazifi (2016).

Given the high pass-through rate of emission costs, especially in carbon-intensive industries, we assume, for simplicity, that the rate is 100% in our calculations. When we discuss the inflation effects on individual consumption goods, we also discuss the effect of a potentially lower pass-through rate in sectors with greater competition.

We assume that the price is implemented without any additional changes in taxation. In some countries other taxes are cut at the same time as environmental taxes are introduced such that the overall taxation level remains the same. The Australian carbon tax implemented in 2012 and the Swedish carbon tax introduced in the early 1990s are two examples of instances where other taxes were reduced to offset the effect of a carbon tax (e.g., Sterner, 1994; Crowley, 2017). A reduction in other taxation may in some cases reduce the upward pressure on aggregate prices when a carbon price is introduced. How much the inflation effect is offset depends on the construction of the tax reform.

Data

To estimate the inflation effect of a carbon price, we must first estimate how much CO_2 each unit of consumption contains. We do so by using input–output tables from the World Input Output Database (WIOD).[4] The WIOD

---

[3] Using a theoretical model, Smale et al. (2006) found that a carbon price would lead to a near-100% pass-through rate in carbon-intensive industries.

[4] See Boitier (2012), Timmer et al. (2015), and Andersson (2018) for a detailed description of the database.

is one of the largest input–output databases available that also includes environmental accounts. It includes harmonized data covering 35 industrial sectors from 40 major economies (see Appendix A for a list of sectors and countries). All other countries (mostly developing countries) are combined into one component: "the rest of the world." The final consumption is calculated from three main groups: households, the government, and investments in fixed capital (i.e., machinery and buildings). It is also possible to estimate exports and imports for a country.

The WIOD contains annual data from 1995 to 2009. In our estimations, we primarily used data from 2008 to estimate the inflation effects because 2009 is affected by the international financial crisis. Most countries observed large declines in output and trade in 2009 as a result of the crisis, which may bias our results. Consequently, we used the most recent year that is not affected by the crisis. To explore how changes in technology and consumption patterns over time impacts our estimates, we also briefly discuss what the inflation effects would have been based on data from 1995 and 2002.

Estimating the Price Effects of a Global Carbon Price

We estimated the inflation effect of a carbon price in two steps. First, we calculated the amount of CO_2 that is contained in each unit of consumption, investments, exports, and imports using the input–output tables. In the second step, we applied the carbon price to the amount of CO_2 emitted to produce each unit of consumption. We thereby obtained an estimate of how much the cost of consumption increases with the carbon price.

The CO_2 content in each unit of final consumption was obtained as follows (see Boitier, 2012 and Andersson, 2015, for the full details). Let x_m be a vector with the total output from country $m = 1,..., N$. The economy is made up of 35 industries, and each element in x_m represents the output from one specific industry. The output is used either for final consumption or as an intermediary good in the production of a final good. The final consumption can be private consumption, investments, or government consumption.

Let $f_{v,m,k}$ represent the final consumption in country m that is produced in country $v = 1,\ldots, N$ and consumed by consumer $k = \{consumer, investments, government\}$. The final consumption can either be private consumption, government consumption, or investments. The total output is equal to:

$$\begin{pmatrix} x_1 \\ \vdots \\ x_m \\ \vdots \\ x_N \end{pmatrix} = \begin{pmatrix} A_{11} & \cdots & A_{1v} & \cdots & A_{1N} \\ \vdots & \ddots & \vdots & \ddots & \vdots \\ A_{M1} & \cdots & A_{mv} & \cdots & A_{mN} \\ \vdots & \ddots & \vdots & \ddots & \vdots \\ A_{N1} & \cdots & A_{Nv} & \cdots & A_{NN} \end{pmatrix} \begin{pmatrix} x_1 \\ \vdots \\ x_m \\ \vdots \\ x_N \end{pmatrix} + \sum_{m=1}^{N} \begin{pmatrix} f_{1,m,k} \\ \vdots \\ f_{v,m,k} \\ \vdots \\ f_{N,m,k} \end{pmatrix} \quad (1)$$

where $A_{m,v}$ is the inter-industrial matrix showing how many intermediate goods country v imports from country m to produce one unit of output. This matrix captures the trade links among countries and allows for the estimation of how much CO_2 each unit of final consumption contains as well as the country where it was emitted.

Thus, (1) can be written as:

$$x = Ax + \sum_m f_m. \quad (2)$$

The output required to produce one unit of final consumption is given by:

$$x = \sum_m (1 - A)^{-1} f_m = \sum_m y_m. \quad (3)$$

Let e_m be a vector of the CO_2 emissions in tons per unit of output in country m. The amount of emissions in country m used to produce one unit of consumption in country v is given by:

$$E_{m,v,k} = e_m y_{m,v,k}, \tag{5}$$

where $y_{m,v,k} = (1 - A_{m,v})^{-1} f_{m,v,k}$.

Having obtained the emission content in each unit of final consumption, we multiplied the emission by the carbon price to obtain the cost of the carbon. We then divided this cost by the value of the final consumption. This gave us an estimate of the inflation effect:

$$PE_{k,v} = \frac{\sum_{m=1}^{41} PC_m \times E_{m,v,k}}{\sum_{m=1}^{41} f_{m,v,k}}, \tag{6}$$

where PC_m is the carbon price in country m, i.e., the country where the emissions originated.

EMPIRICAL RESULTS

The estimated inflation effects for Scenario 1 are shown in Table 1. The inflation effects are modest, despite assuming that there is no change in either behavior or technology. For developed countries, consumer, government, and investment inflation is limited to between 1% and 3.5%. Central banks, such as the Federal Reserve and the European Central Banks, have inflation targets of 2%. Thus, our estimates correspond to one year of normal price increases. The estimated inflation effects are in line with the results of Nong and Siriwardana (2017), who found that a 16 USD/ton CO_2 global carbon price would cause an increase in consumer prices in the European Union of 0.36%, which corresponds to an increase of 2.3% at a 100 USD/ton CO_2 carbon price. These results are also close to the Australian treasury's estimate of a 0.7% increase in consumer prices when the Australian carbon price of 23 AUD/ton CO_2 was introduced in 2012 (Besley et al., 2014).

Table 1. Estimated inflation (%) effect of a global price on carbon of 100 USD/ton CO₂ in 2008

	Consumer	Government	Investment	Export	Import
	100 USD/ton CO2				
European Union	2.3	1.1	2.2	3.9	16.1
United States	3.4	2.5	3.1	9.2	11.6
Other developed countries	2.8	1.7	3.2	10.1	7.2
China	8.2	7.3	13.3	12.0	9.9
India	7.9	6.3	11.4	17.3	14.3
	100 USD/ton CO2—PPP adjusted				
European Union	2.2	1.0	2.0	3.9	9.2
United States	3.2	2.4	2.7	9.2	7.7
Other developed countries	2.6	1.6	2.9	10.1	4.9
China	3.9	3.4	6.3	5.4	9.3
India	2.4	1.9	3.7	4.9	9.7

For developing countries, the inflation effect is double in size, or even three times as high compared to developed countries. Consumer prices in both China and India would increase by approximately 8%. Investment prices would see double-digit increases of between 11% and 13%. This relatively large increase in investment prices is likely to have a significant and negative effect on capital investments. Capital investments is an important component in economic growth. Large increases in investment prices thus represent a potential problem for developing countries with an ambition to grow their economies.

Foreign trade in all countries, developed and developing, would be relatively heavily affected by a global carbon price. Foreign trade is still dominated by trade in goods, and not by trade in services, whereby export and import prices are more affected than are consumer, government, and investment prices: they are between 9% and 16% for all economic regions, except for the EU, where export inflation is limited to 3.9%. The lower export inflation for the EU is explained by below-average carbon intensities and a relatively high share of service exports. India, and especially China, have in recent decades pursued an export-led growth strategy. The carbon price would partly put an end to that strategy.

The conclusion for Scenario 1 is that the inflation effects of a global carbon price are small in developed countries, even under highly unrealistic assumptions that are likely to lead to an overestimation of the inflation effect. If the price is phased in over two or three years, any potentially negative welfare effect is likely to be very small, even in the short run. Moreover, the need for consumers and firms to adjust their consumption patterns and technology to limit the inflation effect is limited, which is likely to make a global carbon price politically feasible.

A clear concern, however, is the change in trade patterns that the carbon price may cause, as export prices from China and India, in particular, will rise by far more than will those from developed countries. Also, the large increase in investment prices is a clear concern. Less exports and less investments will hurt these countries' ability to become richer. The purpose of the carbon price is, of course, to penalize countries with carbon-intensive production methods, but the ability to reduce emissions in the short run is likely to be limited, whereby it is unlikely that these countries will agree to a common carbon price. A simple method to accommodate such concerns in the short run is to temporarily reduce the carbon price for developing countries, which leads us to the second scenario.

In Scenario 2, we weighted the carbon price by the price level in each country, such that the carbon price was made equal across countries in purchasing power terms. Unsurprisingly, adjusting the global carbon price to these differences in price level lowers the price and inflation effect paid by China and India (see bottom half of Table 1). In China, the consumer price inflation effect is reduced from 8.2% to 3.9%, the government consumption inflation effect is reduced from 7.3% to 3.4%, and the effect on investment prices is reduced from 13.3% to 6.3%. For India, the corresponding inflation reductions are from 7.9% to 2.4% (consumer), 6.3% to 1.9% (government), and 11.4% to 3.7% (investments). Overall, the inflation effects are thus reduced to a level similar to that observed in developed countries. As part of the differences in price levels is caused by differences in productivity, it is unsurprising that the inflation effects are more or less equalized when the price is adjusted for differences in price levels.

From a political point of view, it is more likely that a carbon price will be agreed upon to internationally if no country is more negatively affected by it than another country. On the other hand, from a global warming point of view, allowing countries such as China to continue to emit at a reduced cost could increase worldwide emission levels (e.g., Kriegler et al., 2015)—not only because India and China would be allowed to continue to use carbon-inefficient production technologies, but also because some carbon-intensive production could be outsourced from developed countries (Cole, 2004; Andersson, 2018). Over time, as China and India develop and increase their productivity, the carbon prices would increase in these countries. A productivity weighting of the price is thus a "temporary" measure to ensure that the economic burden of a carbon price does not fall too heavily on poorer countries.

In Scenario 3, we assume that all countries are as carbon-efficient as the US (i.e., have the same CO_2/unit of output). Because developing countries are less carbon-efficient, a simple way to reduce emissions is to adopt already existing technologies. The results are presented in Table 2. The top half of the table presents the results for a uniform global price of 100 USD/ton CO_2, and the bottom half of the table shows the results for the price level-adjusted carbon price. What is notable from the results is that the Chinese and Indian consumption patterns are not more carbon-intensive compared to those of developed countries. If China and India adopted the same technologies as the developed countries, the inflation effects would be of a similar magnitude in all countries, between 1% and 3%, depending on the country and consumer group. In this case, there is no need for a price level adjustment. The inflation effects are also so small, even with our unrealistic assumptions, that the implementation of a carbon price of more than 100 USD/ton CO_2 would probably be politically feasible.

Next we consider (i) how the inflation effects have changed over time, by considering what the inflation effects would have been in 1995 and 2002 (the carbon price is adjusted for inflation) given the technology and consumption patterns that existed at that time, and (ii) for consumers, which consumer items would be the most affected by a global carbon price by decomposing consumer price inflation into the consumption groups of food, textiles and clothes, other goods, transport, utilities, and services.

Table 2. Price effects of a global price on carbon of 100 USD/ton CO_2 in 2008, assuming that all countries have the same carbon efficiency as the United States

	Consumer	Government	Investment	Export	Import
100 USD/ton CO2					
European Union	2.8	0.8	0.8	2.0	1.1
United States	1.8	0.9	0.5	1.0	1.0
Other developed countries	2.1	0.8	0.5	0.7	1.9
China	1.7	0.7	0.5	1.3	0.6
India	1.7	2.0	0.6	1.1	0.9
100 USD/ton CO2—PPP adjusted					
European Union	2.8	0.8	0.8	2.0	0.9
United States	1.8	0.9	0.5	1.0	0.9
Other developed countries	2.1	0.8	0.5	0.7	1.8
China	0.7	0.3	0.3	0.6	0.6
India	0.5	0.6	0.2	0.3	0.9

Figure 1 shows the results over time for a uniform carbon price, and Figure 2 shows the results for a price level-adjusted carbon price (i.e., Scenario 2). The most notable change over time is for China, which has become much less carbon-intensive than it was in 1995. A uniform carbon price in 1995 would have increased consumer prices by 40%, as compared to the estimated increase of 8.2% in 2008. Similar effects are shown for government, investment, and export prices. For developed countries, the inflation effect is also estimated to be smaller compared to 1995, but the reduction is only approximately 1% to 1.5%. Nevertheless, this represents a reduction of the inflation effect by between one quarter and one third over a 13-year period. Economic development is clearly important to reduce the relative (although not the absolute) carbon intensity of the economy. Further economic development over time is likely to lead to further reductions in the inflation effects for both China and India.

The price level-adjusted numbers in Figure 2 clearly reduce the inflation impacts in China and India. The reduction is larger in 1995 compared to 2001, when China and India were less developed compared to 2008. This result illustrates that a price level adjustment of the carbon price will lead to increasingly smaller benefits for developed countries as they develop and improve their technology. The price level adjustment would end completely when the developing countries have fully caught up with the developed world.

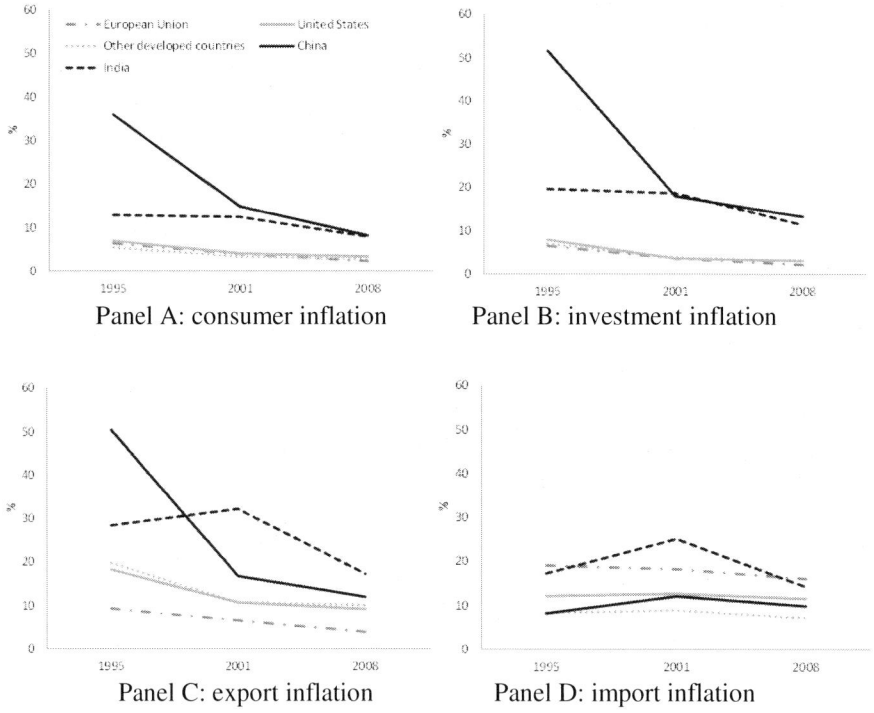

Figure 1. Inflation effects of a global 100 USD/ton CO_2 carbon price.

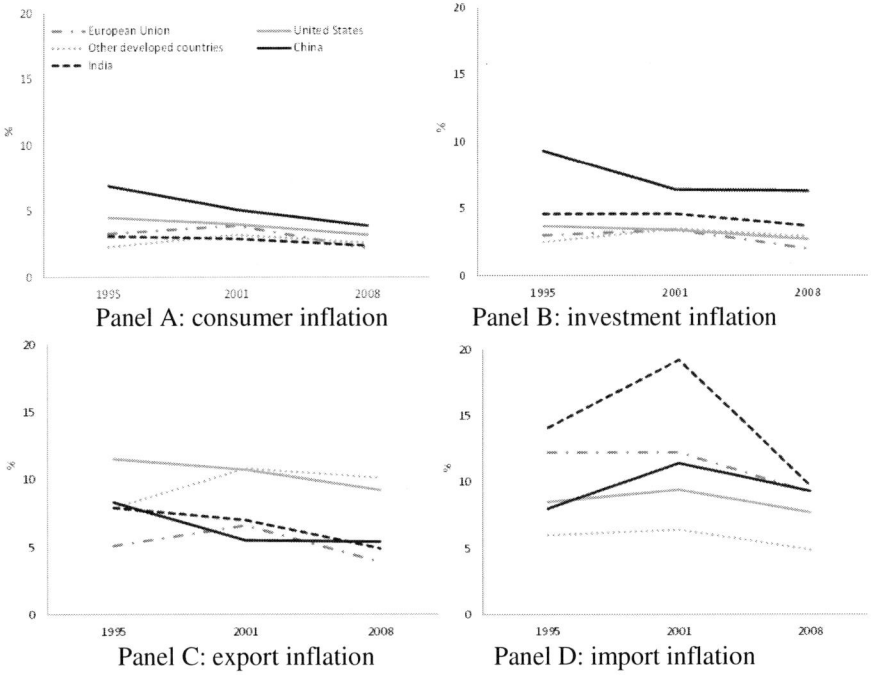

Figure 2. Inflation effects of a purchasing power-adjusted 100 USD/ton CO_2 carbon price.

Finally, we decompose the consumer price effect into different consumer goods. A concern is that a carbon price may hit the poor more than the rich through a greater impact on goods items such as food and clothing. For all countries, the effect on food, clothes, and services are minimal—between 0.3% and 2.3%—when a uniform carbon price is applied, see Table 3. For these items, a price level-adjusted carbon price is unnecessary, as the inflation effects are similar across all countries with a uniform price. Food inflation in China and India, for example, is 1.0% and 2.1%, respectively. In the EU and the US, the corresponding inflation effects are 0.7% and 1.2%, respectively. For the poorest in society, which spend a relatively high share on food, the inflation effects are limited. The inflation effects are also limited for textiles and clothing (between 1.1% and 2.3%).

Table 3. Consumer price inflation with a global price on carbon of 100 USD/ton CO_2

	Food	Textiles and clothes	Goods	Transport	Utilities	Services
100 USD/ton CO_2						
European Union	0.7	1.1	3.1	8.2	29.2	0.3
United States	1.2	2.0	5.1	14.0	90.3	0.6
Other developed countries	0.6	1.5	4.8	7.4	43.4	0.6
China	1.0	1.6	15.5	21.1	364	0.9
India	2.1	2.3	11.7	3.5	460	0.5
100 USD/ton CO_2—PPP adjusted						
European Union	0.7	0.7	2.8	8.0	26.9	0.3
United States	1.2	1.4	4.7	13.7	86.6	0.6
Other developed countries	0.6	1.0	4.3	7.0	40.0	0.6
China	0.5	0.7	7.7	11.1	168	0.4
India	0.6	0.7	4.0	1.2	135	0.2

For other goods and transport there is greater heterogeneity across countries, but it is not necessarily just developing countries that would observe the greatest inflation effects. The second-highest inflation effect on transport is observed for the US, while India has the lowest estimated inflation effect.

The largest difference by far in price effects among the countries is for utilities, where the inflation effect ranges from 29% (EU) to 460% (India). For utilities, electricity generation through coal, in particular, leads to high inflation effects not only for China and India but also for the US compared to the EU. There are two key characteristics of utilities that need to be considered. First, they take time to build and commonly have a relatively long life-span. Second, in many countries, they are either government-owned or at least heavily regulated. In China, utilities are almost entirely owned by the state, and in India only 16% of electricity generation infrastructure is privately owned (Shukla and Tampy, 2011). Both countries also rely heavily on coal for electricity generation. In China, coal prices used in the production of electricity are often subsidized and set below the market value (Zhang, 2014).

The high dependence on state-owned firms in electricity generation in China and India implies that the choice of technology is a government decision rather than a market decision. This has two important implications. Frist, the pass-through rate is likely to be high, as the firms in the sector face no or little competition. Our assumption of a 100% pass-through rate is, thus, likely to be accurate at least for this sector. Second, the market is a key mechanism for reducing emission intensities. Without competition, there is more waste and energy use. A carbon price may reduce energy waste from the consumer's side but not the producer's side, as the producer can pass on the entire cost to the consumer. Thus, the carbon price will not automatically lead to new technologies and energy sources being introduced. Moreover, the carbon price, in itself, is thus an insufficient policy to reduce emissions and limit their impact. The carbon price must be accompanied by other policies aimed at the state-owned firms in the electricity (utilities) sector.

So far, we have assumed a 100% pass-through rate. Given previous studies, this rate is realistic for utilities where most of the inflation effects take place. For other sectors with greater competition, such as food and textiles, the pass-through rate is potentially smaller. A lower pass-through rate will lead to a lower inflation rate. Given that the inflation effects on prices, except for transport and utilities, are so small, the welfare losses, even with a 100% pass-through rate, are so small that the carbon price is unlikely to have any major negative welfare effects. The conclusion is that it is already possible to impose a relatively large carbon price, given existing technologies. The only problem is the utilities sector and, potentially, the transport sector in developing countries such as China and India.

CONCLUSION

This chapter maps the potential effect of a global carbon price on aggregate prices in the economy. Our results show that the inflation effect is modest in developed countries: between 1% and 3% for consumer prices, government prices, and investment prices. These price increases correspond to one year of normal inflation. The inflation effects are higher in India and China; around 7% to 8%. Our results also show that a global carbon price would affect countries' international competitiveness. The European Union has the most to gain, while China and India have the most to lose. In addition, investment prices would increase by double digits in the latter two countries, which would limit firm investments and long-term economic growth. It is highly unlikely that China and India will agree to a uniform carbon price given these results.

For developed countries, the inflation effects are mostly negligible. This is not the case for developing countries such as China and India. It will be politically difficult for developing countries to introduce a globally uniform carbon price. However, a simple method to reduce the inflation effect in China and India is to weight the carbon price based on productivity. A productivity-weighted carbon price reduces the inflation effect in China and India to between 2% and 4%. Over time, as China and India develop their economies, the price would automatically increase in line with their productivity.

The main conclusion of this chapter is thus that a 100 USD/ton CO_2 price is feasible to implement in developed countries but not in developing countries, due to their lower levels of economic development. However, by weighting the price according to productivity (i.e., development), it is more likely that poorer developing countries would agree to introducing a carbon price.

REFERENCES

Ackerman, F., DeCanio, S. J., Howarth, R. B. and Sheeran, K. (2009). Limitations of integrated assessment models of climate change. *Climate Change*, 95: 297-315.

Aldy, J. E. and Stavin, R. N. (2012). The promise and problems of pricing carbon: Theory and experience. *Journal of Environment & Development*, 21(2): 152-180.

Andersson, F. N. G. (2018). International Trade and Carbon Emissions. The Role of Chinese Institutional and Policy Reforms. *Journal of Environmental Management*, 205(1): 29-39.

Balassa, B. (1964). The purchasing power parity doctrine: A reappraisal. *Journal of Political Economy*, 72: 584-596.

Besley, D., Downie, C., Kennedy, S. and Writer, S. (2014). Carbon pricing in Australia: an early view from the inside. In Quiggin, J., Adamson, D., Quiggin, D. (eds). *Carbon Pricing: Experience and future prospects*. Cheltenham: Edward Elgar Publishing Ltd.

Bosquet, B. (2000). Environmental tax reform: does it work? A survey of the empirical evidence. *Ecological Economics*, 34: 19-32.

Böhringer, C. and Rutherford T. F. (2008). Combining bottom-up and top-down. *Energy Economics*, 30: 574-596.

Caballero, R. J. (2010). Macroeconomics after the Crisis: Time do deal with the pretense-of-knowledge syndrome. *Journal of Economic Perspectives*, 24: 85-102.

Carpos, P., Paroussos, L., Fragkos, P., Tsani, S., Boitier, B., Wagner, F., Busch, S., Resch, G., Blesl, M. and Bollen, J. (2014). Description of models and scenarios used to assess European decarbonization pathways. *Energy Strategy Reviews*, 2(3-4): 220-230.

Cole, M. A. (2004). Trade, the pollution haven hypothesis and the environmental Kuznets curve: examining the linkages. *Ecological Economics*, 48(1): 71-81.

Crowley, K. (2017). Up and Down with Climate Politics 2013-2016: The Repeal of Carbon Pricing in Australia. *WIREs Climate Change*, 8: 1-13.

De Bruyn, S. M., Vergeer R, Shep E, Hoen M. T., Korteland M, Cludius J, Schumacher K, Zell-Ziegler, C. and Healy S. (2015). *Ex-post investigation of cost pass-through in the EU ETS. An analysis for six sectors*. European Union Commission, November 2015.

Dennig, F., Budolfson, M. B., Fleurbaey, M., Siebert, A. and Socolow, R. H. (2015). Inequality, climate impacts on the future poor, and carbon prices. *PNAS*, 112(52), 15827-15832.

Dong, D. and Siriwaranda, M. (2017). Environmental and economic impacts of a joint emission trading scheme. *International Journal of Global Energy Issues*, 40 (3/4): 184-206.

Fabra, N. and Reguant, M. (2014). Pass-through of emission costs in electricity markets. *American Economic Review*, 104(9): 2872-2899.

Faber A., Idenburg A. M. and Wilting, H. C. (2007). Exploring techno-economic scenarios in and input-output model. *Futures*, 39: 16-37.

Farmer, J. D., Hepburn, C., Mealy, P., and Teytelboym, A. (2015). A third wave in economics of climate change. *Environmental and Resource Economics*, 65(2): 329-357.

Hurtado, S. (2014). DSGE models and the Lucas critique. *Economic Modeling,* 44: 512-519.

Hourcade, J. C. and Jaccard, M. (2006). Hybrid Models: New answers to old challenges. Introduction to the special issue of the Energy Journal. *Energy Journal*, 27: 1-11.

IPCC (2014). Summary for Policy Makers. In: *Climate Change 2014: Mitigation of Climate Change. Contribution of Working Group III to the Fifth Assessment Report of the Intergovernmental Panel on Climate Change*. Cambridge: Cambridge University Press.

King, M. (2016). *The end of Alchemy. Money, banking and the future of the global economy*. Little, Brown, London.

Kriegler, K. Riahi, N. and Bauer. (2015). Making or breaking climate targets: The AMPERE study on staged accession scenarios for climate policy. *Technological Forecasting and Social Change*, 90(Part A): 24–44.

Laing, T., M. Sato, M. Grubb, C. and Combert (2014). The effects of and side-effects of the EU emissions trading system. *WIREs Climate Change*, 5: 509-519.

Lo, A. Y. (2012). Carbon emissions trading in China. *Nature Climate Change*, 7: 765-766.

Lubik, T. A. and Surico, P. (2010). The Lucas Critique and the Stability of Empirical Models. *Journal of Applied Econometrics*, 25(1):177-194.

Lucas, R. E. (1976). Econometric Policy Evaluation: A Critique. *Carnegie Rochester Conference Series on Public Policy* 1, 19-46.

Meng, S., Siriwardana, M. and McNeill, J. (2013). The environmental and economic impact of the carbon tax in Australia. *Environmental Resource Economics*, 54: 313-332.

Nazifi, F. (2016). The pass-through rates of carbon costs on to the electricity prices within the Australian National Electricity Market. *Environmental Economics and Policy Studies*, 18(1): 41-62.

Newell, R. G., Pizer, W. A. and Raimi, D. (2013). Carbon markets 15 years after Kyoto: Lessons learned, new challenges. *Journal of Economic Perspectives*, 27(1): 123-146.

Oliner, S. D., Rudebusch, G. D. and Sichel D. (1996). The Lucas critique revisited. Assessing the stability of empirical Euler equations for investment. *Journal of Econometrics*, 70: 291-316.

Pearce, D. (1991). The role of carbon taxes in adjusting to global warming. *The Economic Journal*, 101: 938-948.

Peterson, G. D., Cumming, G. S. and Carpenter, S. R. (2003). Scenario planning: A tool for conservation in an uncertain world. *Conservation Biology*, 17(2): 358-366.

Pindyck, R. S. (2013). Climate Change Policy: What do the models tell us? *NBER Working paper*, 19244.

Ravallion, M. (2013). Price levels and economic growth: Making sense of revisions to data on real incomes. *The Review of Income and Wealth*, 59(4): 593-613.

Revesz, R. L., Howard, P. H., Arrow K., Goulder, L. H., Kopp, R. E., Livermore, M. A., Oppenheimer, M. and Sterner, T. (2014) Global warming: improve economic models of climate change. *Nature*, 508: 173–175.

Samuelson, P. (1964). Theoretical notes on trade problems. *Review of Economics and Statistics*, 46: 145-154.

Smale, R., Hartley, M., Hepburn, C., Ward, J. and Grubb, M. (2006). The impact of CO_2 emissions on firm profits and market prices. *Climate Policy*, 6(1): 31-48.

Sterner, T. (1994). Environmental tax reform: The Swedish experience. *Environmental Policy and Governance*, 4(6): 20-25.

Timmer, M. P. (Ed.). (2012). *The World Input-Output Database (WIOD): Contents, Sources and Methods.* World Input-Output Database.

Van Vuuren, Detlef P., Isaac, M., Kundzewics, Z. W., Arnell, N., Barker, T., Criqui, P., Berkhout, F., Hilderink, H., Hinkel, J., Hof, A., Kitous, A., Kram, T., Mechler, R. and Scricieu, S. (2011). The use of scenarios as the basis for combined assessment of climate change mitigation and adaptation. *Global Environmental Change*, 21: 575-591.

Weitzman, M. L. (2014). Can Negotiating a Uniform Carbon Price Help to Internalize the Global Warming Externally? *Journal of the Association of Environmental and Resource Economists*, 1(1): 29-49.

Zhang, Z. X. (2014). Energy prices, subsidies and resource tax reform in China. *Asia Pacific Policy Studies*, 1(3): 439-454.

APPENDIX A

Countries included in the study:

Australia
Austria
Belgium
Bulgaria
Brazil
Canada
China
Cyprus
Czech Republic
Denmark
Estonia
Finland
France
Germany
Greece
Hungary
Indonesia
India
Ireland

Italy
Japan
South Korea
Latvia
Luxembourg
Lithuania
Mexico
Malta
Netherlands
Poland
Portugal
Romania
Russia
Slovakia
Slovenia
Spain
Sweden
Turkey
Taiwan
United Kingdom
United States
Rest of the world

APPENDIX B

Industrial sectors included in the study:

Agriculture, hunting, forestry, and fishing
Mining and quarrying
Food, beverages, and tobacco
Textiles and textile products
Leather and footwear
Wood and products of wood and cork
Pulp, paper, and printing and publishing
Coke, refined petroleum, and nuclear fuel
Chemicals and chemical products
Rubber and plastics
Other non-metallic minerals

Basic metals and fabricated metal
Machinery
Electrical and optical equipment
Transport equipment
Manufacturing
Electricity, gas, and water supply
Construction
Sale, maintenance, and repair of motor vehicles and motorcycles, retail of fuels
Wholesale trade and commission trade, except motor vehicles and motorcycles
Retail trade, except motor vehicles and motorcycles, repair of household goods
Hotels and restaurants
Inland transport
Water transport
Air transport
Other supporting auxiliary transport activities, activities of travel agents
Postal and telecommunications
Financial intermediation
Real estate activities
Rent of M&Eq and other business activities
Public administration, defense, and compulsory social security
Education
Health and social work
Other community, social, and personal services
Private households with employed persons

INDEX

A

adjustment, x, 87, 100, 102
agricultural crops, viii, 42, 49, 55
agricultural sector, 79
agriculture, 43, 59, 71, 78, 79, 80
alkaline earth metals, 25, 28, 31
Amazon forest, viii, 42, 45, 58, 78, 79
amine, vii, viii, 1, 2, 3, 4, 5, 7, 8, 9, 10, 11, 12, 13, 14, 15, 16, 18, 20, 21, 24, 32, 33, 34
ammonia, 24, 25, 28, 31, 33
aqueous amine solution, vii, 1, 2, 5, 7, 8, 9, 11, 12, 14, 15, 16, 21
aqueous solutions, 13, 16, 28, 31, 33
armed forces, 77, 83
assessment, 34, 71, 72, 78, 106, 110
assessment models, 106
Atlantic forest, viii, 42, 44, 45, 51, 52, 54, 56, 57
atmosphere, vii, viii, 1, 2, 25, 41, 42, 45, 46, 47, 49, 64
atmospheric pressure, 2, 47

B

balance sheet, 84
barium, 7, 14, 15
bicarbonate, 3, 10
biochemical processes, viii, 41
biodiesel, 35, 36, 37, 38, 39
biodiversity, 44, 72, 73, 77
biofuel, 53
biological processes, 47
biomass, 45, 48, 55, 59, 60
boreal forest, 61
bottom-up, 90, 107
Brazil, v, vii, viii, ix, 41, 43, 44, 45, 46, 49, 50, 53, 56, 58, 61, 62, 63, 64, 67, 69, 70, 71, 73, 75, 77, 79, 80, 83, 85, 86, 110

C

calcium, 37
carbon, vii, viii, ix, 2, 3, 15, 22, 25, 26, 27, 28, 30, 31, 32, 33, 34, 41, 43, 44, 45, 46, 48, 49, 53, 54, 55, 57, 59, 61, 64, 68, 73,

79, 80, 82, 83, 87, 88, 89, 90, 91, 92, 93, 94, 95, 97, 98, 99, 100, 101, 102, 103, 104, 105, 106, 107, 109
carbon dioxide, 1, iii, 2, 22, 25, 26, 27, 28, 30, 31, 32, 33, 34, 42, 46, 53, 54, 59, 61, 64, 84, 88
carbon emission, v, 67, 68, 73, 79, 82, 107, 108
carbon price, v, vii, ix, 87, 88, 89, 90, 91, 92, 93, 94, 95, 97, 98, 99, 100, 101, 102, 103, 105, 106, 107, 110
carbon pricing, 88, 107
challenges, vii, 1, 2, 69, 108, 109
chemical, vii, viii, 1, 2, 3, 10, 11, 13, 15, 20, 22, 23, 41, 45, 46, 47, 59, 61, 111
chemical absorption, vii, 1, 2, 3, 10, 11, 13, 15, 22, 23
chemical characteristics, viii, 41, 46
chemical properties, 59, 61
China, 43, 68, 69, 88, 89, 93, 98, 99, 100, 101, 102, 103, 104, 105, 106, 108, 110
climate change, vii, ix, 1, 2, 43, 49, 57, 59, 60, 61, 67, 68, 72, 80, 81, 82, 83, 84, 85, 86, 87, 88, 106, 107, 108, 109, 110
CO_2, v, vii, viii, 1, 2, 3, 4, 5, 7, 8, 9, 10, 11, 12, 13, 14, 15, 16, 17, 18, 19, 20, 21, 22, 23, 24, 25, 28, 30, 31, 32, 33, 34, 35, 41, 42, 43, 44, 45, 46, 47, 48, 49, 50, 51, 52, 53, 54, 55, 57, 58, 59, 60, 61, 62, 63, 64, 65, 69, 82, 84, 88, 91, 92, 93, 94, 95, 96, 97, 98, 100, 101, 102, 103, 104, 106, 109
combustion, viii, 25, 28, 30, 41, 43
commercial, 7, 14
community, 46, 49, 112
compaction, 48
competition, 94, 105
compilation, vii
complement, 21, 89, 92
compliance, ix, 68, 79
composition, 31, 43, 72
compounds, 23, 25, 28, 31
conference, 68, 70
consumer goods, 103
consumer prices, vii, ix, 87, 88, 89, 90, 92, 97, 98, 101, 105
consumption, 3, 24, 88, 89, 92, 94, 95, 96, 97, 99, 100
consumption patterns, 89, 92, 95, 99, 100
COP21, v, ix, 67, 68, 69, 70, 73, 79, 86
correlation, viii, 42, 47, 48, 51, 53, 73
cost, 10, 11, 13, 32, 55, 93, 95, 97, 100, 105, 107
crop, 48, 53, 55, 63, 79
crop production, 63
crop residue, 48, 55
crop rotations, 53, 63

D

data analysis, 55
database, 94
decomposition, 46, 47, 48, 51, 92
deforestation, vii, viii, ix, 41, 43, 46, 59, 60, 61, 67, 71, 72, 73, 74, 75, 76, 77, 78, 80, 81, 82, 84
deforestation in the Amazon, 73
degradation, 3, 34, 46
Department of Energy, 32
developed countries, x, 87, 89, 93, 97, 98, 99, 100, 101, 102, 104, 105, 106
distribution, 43, 53, 54, 63, 72, 83
drying, 10, 27, 29, 31
dynamic systems, 61

E

economic activity, 91
economic change, 44
economic development, 101, 106
economic growth, 98, 106, 109
economic theory, 88
economic welfare, 88, 90
economics, 91, 108

ecosystem, 43, 44, 45, 46
electricity, 70, 89, 94, 104, 105, 108, 109
emission, viii, ix, 41, 42, 43, 46, 49, 53, 54, 56, 61, 62, 63, 64, 67, 68, 69, 71, 73, 85, 94, 97, 100, 105, 108
energy, 3, 27, 48, 69, 70, 81, 88, 90, 92, 105
engineering, vii, 1, 2
environment, 49, 82, 86
environmental change, 46
environmental crisis, 75
environmental factors, 51, 63
environmental impact, 69
environmental issues, 75
environmental policy, ix, 68
environmental quality, 45
environmental services, 43, 44
equilibrium, 10, 20, 21, 37, 90
equipment, viii, 2, 27, 32, 112
EU Emissions Trading Scheme, 94
European Union, 43, 88, 89, 97, 98, 101, 104, 106, 107
experimental design, 53
export prices, vii, ix, 87, 88, 92, 99, 101
export-led growth, 98

F

filtration, 10, 23, 26, 29, 31, 32
financial crisis, 95
fires, 72, 75, 77, 80, 84, 85
flue gas, viii, 2, 4, 22, 25, 28, 30, 31, 34
food, 36, 59, 92, 101, 103, 105
food security, 59
forest ecosystem, 58
forest formations, 44
forest management, ix, 68, 79
forest restoration, 43

G

global climate change, 88

global economy, 108
global warming, 43, 68, 70, 81, 82, 83, 100, 109
greenhouse gas, ix, 32, 33, 34, 42, 56, 62, 63, 65, 67, 68, 69, 70, 71, 76, 79
greenhouse gas emissions, 62, 68, 70, 71, 79
greenhouse gases, 42

H

heterogeneity, 45, 47, 55, 104
humidity, 46, 47, 49, 55
hydrocarbons, 25, 28, 30
hydroxide, 7, 14, 24
hypothesis, 107

I

import prices, v, vii, ix, 87, 88, 92, 98
India, 69, 89, 93, 98, 99, 100, 101, 102, 103, 104, 105, 106, 110
industrial revolution, 42
industrial sectors, 95
industries, 94, 95
inflation, v, vii, ix, 87, 89, 92, 94, 95, 97, 98, 99, 100, 101, 102, 103, 104, 105, 106
integration, 21, 24, 27, 34, 79
international competitiveness, 106
investment, vii, ix, 24, 32, 70, 87, 88, 90, 92, 97, 98, 99, 101, 102, 103, 106, 109
investment prices, vii, ix, 87, 88, 90, 92, 98, 99, 106
investments, 70, 95, 96, 98, 99, 106

L

light, x, 44, 45, 48, 87
liquid phase, 21, 27, 29, 31

M

management, ix, 42, 49, 54, 55, 56, 63, 68
measurement, 51, 60, 85
metals, 25, 28, 31, 94, 112
mineralization, 48
models, 21, 43, 72, 89, 90, 91, 92, 107, 108, 109
moisture, viii, 42, 46, 47, 48, 50, 51, 52, 53, 54
molecular weight, 8
multivariate analysis, 55

N

nanoparticles, 36, 37
native species, 43
nitrates, 25, 26, 28, 30, 31
nitrification, 57
nitrogen, 22, 23, 24, 25, 28, 30, 44, 48, 55, 56, 57
nitrous oxide, 42, 59

O

oil, 36, 39, 70, 86
operating costs, 24, 32
operations, 22, 53, 54, 63
opportunities, 34, 45
optimization, vii, viii, 1, 62
organic compounds, 25, 28, 31
organic matter, 45, 46, 47, 48, 49, 51, 58
oxidation, 25, 28, 30, 49
oxygen, 25, 28, 30, 32, 34, 77
ozone, 22, 23, 25, 29, 31, 32

P

Paris agreement, vii, ix, 68, 69, 72, 82, 86
phosphorus, 46, 48
physical characteristics, 46
physical properties, 47
physical structure, viii, 41, 47
policy, ix, 67, 71, 72, 85, 88, 89, 90, 91, 93, 105, 108
porosity, viii, 41, 48, 55, 64
positive correlation, 47, 48, 55
power generation, 34
power plants, 2
precipitation, 3, 7, 9, 10, 14, 15, 16, 18, 19, 20, 24, 25, 28, 30, 51
price effect, 88, 89, 93, 103, 104
purchasing power, 93, 99, 103, 107
purchasing power parity, 107

R

rain forest, 58, 59, 61
rainfall, 49, 51, 53, 70
rainforest, 45, 46, 50, 54, 60, 64, 84
regeneration, 3, 22, 27, 32, 51, 57
respiration, viii, 41, 42, 45, 46, 47, 48, 51, 54, 55, 56, 58, 60, 61
restoration, 45, 62, 72, 79, 81

S

scale-up, viii, 2, 13, 20, 21, 33
services, iv, 45, 92, 98, 101, 103, 112
simultaneous removal, viii, 2, 22, 24, 25, 28, 30
sodium hydroxide, 7, 14, 31
soil CO_2 emission, v, viii, 41, 42, 46, 49, 57, 62, 63, 64
soil respiration, 42, 45, 48, 54, 55, 56, 58, 60, 61
solid phase, 26, 27, 29, 31
solution, 2, 3, 5, 7, 8, 9, 13, 14, 15, 16, 18, 19, 20, 21, 22, 25, 27, 28, 31, 32, 34, 72
stability, 10, 36, 38, 48, 109
state, ix, 17, 21, 59, 68, 78, 79, 104, 105

sugarcane, ix, 42, 51, 53, 54, 57, 62, 63, 64
sulfur, 22, 23, 24, 28, 30, 34, 48
sulfur dioxide, 34
surface area, 44
surveillance, 78
sustainable development, 79
syndrome, 107

United Kingdom, 111
United Nations, ix, 42, 43, 59, 67, 70, 81, 85, 86
United Nations Framework Convention on Climate Change (UNFCCC), ix, 67
United States, 43, 89, 98, 101, 104, 111
urea, 23, 25, 28, 31

T

tax reform, 89, 94, 107, 109, 110
tax reforms, 89
taxation, 94
taxes, 94, 109
techniques, 65
technological change, 89
telecommunications, 112
temperature, 2, 16, 20, 25, 27, 28, 30, 46, 47, 48, 49, 51, 52, 54, 55, 61, 69, 71, 91
temporal variation, 48, 54, 55, 58, 60, 65
trade, 85, 95, 96, 98, 99, 109, 112
transport, viii, 41, 47, 48, 101, 104, 105, 112
tropical forests, 44, 72, 80, 83, 85

V

variables, 3, 20, 34, 56, 72
variations, 54, 55, 70
vegetation, 42, 44, 45, 46, 47, 49, 63, 72, 77, 78, 79

W

waste, 37, 48, 88, 105
water, 8, 37, 44, 45, 48, 54, 55, 70, 112
welfare, 99, 105
welfare loss, 105
wool, 5
worldwide, 77, 92, 100

U

uniform, 100, 101, 103, 106

Y

young people, 82